Electrification of Particulates in Industrial and Natural Multiphase flows

D1808844

Zhaolin Gu · Wei Wei

Electrification of Particulates in Industrial and Natural Multiphase flows

 Springer

Zhaolin Gu
School of Human Settlements
  and Civil Engineering
Xi'an Jiaotong University
Xi'an, Shaanxi, China

Wei Wei
School of Energy and Power Engineering
Wuhan University of Technology
Wuhan, Hubei, China

ISBN 978-981-10-9767-6   ISBN 978-981-10-3026-0   (eBook)
DOI 10.1007/978-981-10-3026-0

Printed on acid-free paper

This Springer imprint is published by Springer Nature
The registered company is Springer Nature Singapore Pte Ltd.
The registered company address is: 152 Beach Road, #21-01/04 Gateway East, Singapore 189721, Singapore

Preface

The understanding of electrical phenomenon is a long-term journey in human history. Electrostatics is an important branch of modern physics. Currently, the applications of electrostatic charge in technological fields are growing, such as electrostatic precipitation, electrostatic spraying, electrostatic separation, and electrostatic printing. Surprisingly, despite being so well-known, the charging mechanism remains poorly understood, and even the most basic questions are still being debated, especially in the particulate entrained fluid systems with identical physical properties. For example, how does the external fluid flow affect particulate charging? What causes the particulate charging? What species are transferred between surfaces to generate the net charge? What is the driving force for charge transfer? Does humidity (i.e., water) or other adsorbed species play a role in the charging? These problems need to be solved imminently and comprehensively.

It is well-known that particulate entrained in natural and industrial flows has two basic phases – liquid phase and solid phase. Droplets can be charged under the preset/associated electric field. Three effective methods can make droplets charged in industrial applications: corona charging, contact charging, and induction charging. In comparison, particle charging through collisions could be attributed to electron transfer, ion transfer, material transfer, and/or aqueous ion shift on particle surfaces. The charging/charged particles would produce self-excited electric field. A review on electrification of particulate entrained fluid flows – mechanisms, applications, and numerical methodology – was written by invitation, published in *Physics Reports: A Review Section of Physics Letters*, 600: 1-53, 2015. The material for the book has evolved largely from this review. Different charging ways in gas (liquid)-liquid or gas-solid multiphase flows are summarized in this book. The computational fluid dynamics method is introduced to study characteristics of particulate electrification in multiphase flow.

Scope

The purpose of this book is to offer theoretical references to control and utilize the charging/charged particulate entrained fluid system, present an extended review of mechanisms and basic theories as well as potential applications of particulate electrification, and provide the research methodologies of particulate charging processes in multiphase flows.

The first chapter of the book provides general description on the electrification of particulate in industrial process and natural phenomenon. Chapter 2 describes the particulate properties in multiphase flows, the properties including particulate forms, particulate-fluid interaction, particulate-particulate interaction, and the mechanical behavior under the electric field.

For the liquid particulate, the charged ways of droplet and basic theories entrained fluid flows are classified in Chap. 3. In this chapter, we consider the droplet would be charged by six different ways, and three charging theories can explain the droplet charging. Most of the charged droplet flow systems are involved in the dynamics of two phases which would affect the charging process; the numerical methodologies considering the droplet entrained fluid flows to study the characteristics of droplet electrification of two-phase flows are presented in Chap. 4. For charged droplet, the numerical methodology would give numerical solutions of the whole electrospray process, not only the electricity characteristic but also the cone-jet formation, jet breakup, charged droplet generation, etc.

For the solid particulate, the charged ways of particle and basic theories entrained fluid flows are classified in Chap. 5. The particle would be charged due to electron transfer, ion transfer, material transfer, and aqueous ion shift on particle surface. The charging models which describe the particle contact electrification are also reviewed in this chapter. The numerical methodology to study the characteristics of particle electrification of two-phase flows is presented in Chap. 6. Combining with the particle charging models, the numerical methodology has the abilities to monitor the particle charging process in particle entrained fluid flows.

Chapter 7 introduces the experimental methods for particle charging; the measurement principles of electrostatic potential, charge quantity, charge density, and powder charging are presented in this chapter. Electrostatic utilization and protection are summarized in Chap. 8. The main contents and general measures in particulate electrification are introduced in this chapter.

At the end of this book, the potential applications of particulate electrification are drawn in Chap. 9.

Xi'an, Shaanxi, China   Zhaolin Gu
Wuhan, Hubei, China   Wei Wei

Acknowledgments

Many people have contributed to the development of this work. Dr. Jian Li from Springer provided the initial encouragement to write the book. A sincere depth of gratitude is extended to Mr. Sen Wang who played an indispensable role in polishing the manuscript to improve the readability. Professor Gu is also indebted to his wife, Hui Zhou, for her continual support and encouragement.

Professor Gu acknowledges the assistance and contributing research of former students in his group. These people include Dr. Yongzhi Zhao, Associate Professor of Zhejiang University; Dr. Jian Qiu, Chinese Academy of Sciences; Dr. Luyi Lu, Associate Professor of Huazhong University of Science and Technology; and Dr. Junwei Su, Associate Professor of Xi'an Jiaotong University.

These assistance and contributing research were continually supported by the National Natural Science Foundation of China (nos. 40675011, 10872159, and 11302155) and the Key Grant Project of the Chinese Ministry of Education (no. 708081).

Contents

About the Authors

Zhaolin Gu is Professor of environmental fluid dynamics at Xi'an Jiaotong University (XJTU), China. He received his BS degree in chemical equipment and machinery from XJTU in 1987, his MS degree in process equipment engineering from XJTU in 1990, and his PhD degree in power engineering and engineering thermo-physics from XJTU in 1997. He did his postdoctoral study in the School of Science and Technology of Keio University, Japan, from 1999 to 2001. He was an Academic Visitor at the Department of Chemical Engineering of Imperial College London, UK, from March to August, 2005. Besides, he was awarded the 2010 and 2017 JSPS Invitation Fellowship for Research in Japan (short-term, 2 months) by the Japan Society for the Promotion of Science (JSPS). He joined XJTU as a teaching assistant in 1990 and got his Associate Professor position in 1996 and his Professor position in 2001. In XJTU, he taught thermal fluid sciences, numerical modeling, and multiple-phase flows. His research area of emphasis is the multiple-phase flows in chemical engineering and environmental science. He has coauthored a book titled *Large Eddy Simulation of Wind Flows in Urban Built Environment*, Science Press, Beijing, 2014 (in Chinese). He has also published a book titled *Wind-Blown Sand: Near-Surface Turbulence and Gas-Solid Two-Phase Flow*, Science Press, Beijing, 2010 (in Chinese).

Wei Wei is Associate Professor of the Energy and Power Engineering School at Wuhan University of Technology. He got his PhD degree in mechanical engineering from XJTU in 2012, supervised by Prof. Gu.

Based on the fact that water content is universal but usually a minor component in most particle systems, Dr. Gu's team proposed a charging mechanism of sands with the adsorbed water on microporous surface in wind-blown sand system and developed the CFD (computational fluid dynamics)-DEM (discrete element method) to demonstrate the dynamics of the sand charging. The charging mechanism could provide an explanation for the charging process of all identical granular systems with water content, including Martian dust devils, wind-blown sands, and even powder electrification in industrial processes. Dr. Gu's team also developed the volume-of-fluid (VOF) method, with fluid flow equation and charge conservation equation, to investigate the deformation and mechanical behavior of leaky dielectric droplet and charged droplet in the external electric field, and proposed a physical model of electric field induced by charged droplets to describe the effect of space charged droplets in electrostatic spray technology.

Part I
Particulate Charging Phenomena in Multiphase Flows

Chapter 1
Electrification of Particulate in Multiphase Flows

1.1 Electrification of Particulate in Industrial Processes

As modern industry develops by leaps and bounds, static electricity is found as hazard in many industrial sectors. Actual conditions of modern industrial production indicate that product material and equipment can generate and accumulate copious static electricity during the process. Electrostatic force and electrostatic discharge can cause damages to industrial production. Electrostatic hazards in-process of industrial production, transport and storage can generally be divided into two categories. One is sudden blast accidents – electrostatic field generates electrostatic discharge, which detonates surrounding explosive air mixture. As what happened in 1969, three 200,000-ton oil tankers owned by Netherlands and other countries were detonated by electrostatic fire, two of which sunk and caused many casualties in less than 3 weeks. According to statistics, a significant proportion of 243 dust explosions happened during 1900~1978 in the United States were electrostatic accidents. The other one is that although electrostatic field is not strong enough to discharge, adsorption and repulsive effect generated by electrostatic force can still bring troubles to industrial production and daily life. For instance, when transporting powder and liquid through pipelines, the electrostatic resistance can cause energy loss; when sieving powder and liquid, the plugging can disrupt productivity; electrostatic force can also cause scouring and bleaching of silks, bonding and winding problems that adversely impact the producing process in fiber production. The former electrostatic hazard is usually disastrous and leads to massive one-time loss, hence it is called electrostatic disaster. Most of electrostatic disasters occur in industrial transportation sectors like oil industry, chemical industry, oil tankers, aircraft and initiating explosive devices. Although the latter electrostatic hazard is not as severe as the former one, it is widely distributed and carries a lot of clout; therefore, the inconvenience it brings to industrial production and daily life shall not be under-estimated. People call such electrostatic hazard as static trouble.

© Springer Nature Singapore Pte Ltd. 2017
Z. Gu, W. Wei, *Electrification of Particulates in Industrial and Natural Multiphase flows*,
DOI 10.1007/978-981-10-3026-0_1

1.1.1   Electrification in Petroleum Industry

Petroleum products are poor conductors of electricity. Therefore, static electricity may be discharged when petroleum products are produced, stored, fueled, and used. Static electricity is dangerous for these products because static electricity sparks may cause explosion or fires. In recent years, researchers at home and abroad have summed up lessons learned from the accidents. Upon a series of researches, they have proposed multiple effective security measures. However, petroleum-related accidents caused by static electricity have never been eradicated because of the complex mechanism of and factors involved with static electricity, various sudden interference or accidental factors in particular. Thus, we must continue to put a premium on safety related to petroleum product and guard against static electricity.

① Flow Electrification Phenomenon

There is an electric double layer at the contact interface of oils and oil pipes due to the transfer of positive and negative ions. It is generally believed that the electric charges of liquid at the solid surface are composed of two parts: one part is the electric charge layer clinging to the solid surface, which is called compact layer with a thickness equivalent to the order of magnitude of a molecular diameter, the electric charges of this layer are of opposite sign with that on the solid surface; the other part is called diffusion layer, in which the electric charges are of the same sign with that in compact layer, the thickness of diffusion layer is dozens to hundreds of times of molecular diameter.

When oils with low conductivity flow through the pipes, there will generate positive charge above the solid interface as well as in the diffusion layer of the fluid while negative charge will be found in the fixed layer. The positive charge forms liquid flow current or streaming current as the liquid flows. The numerical value of streaming current is equal to the volume of the electric power that passes through the tube cross section in a cycle time. If the pipes are made of insulating material or insulation to ground, then there will accumulate static electricity inside the pipes; the quantity of it depends on multiple factors: the flow rate and conductivity of oils, pipe diameter, etc.

② Electrical discharge phenomenon of gas phase space

When oils in oil fuel tanks are electrified, they will act electrically towards the oils themselves or gas phase space inside the tanks. Usually, electrical discharge inside gas phase space is more likely to cause disaster than that inside the oils. When oils carry electric charge with density of $10^{-6} \sim 10^{-5}$ C/m, there will generate an electric field which makes air and other gas molecules ionize and then become ions. Discharge is the ionization of gas molecules; ions become carriers and make the ionization space transform from insulator to conductor with current flows. Therefore, discharge will cause insulation breakdown; electric-field strength of which is called the insulation breakdown field strength (30 kV/cm for air). Besides, discharge can also generate blue-white light and disruptive noise, as well as radiating electromagnetic waves. In other words, if discharge energy of charged oils exceeds the ignition energy of mixed inflammable gases, it may become the ignition source and trigger on explosion.

Discharge in oil fuel tanks can be generally divided into three categories according to its lighting form: corona discharge, brush discharge and spark discharge.

1.1.2 Electrification in Shipping Industry

The most basic electrification in shipping industry is contact electrification.

① Spray electrification of tank cleaning for oil tankers and charged liquid particulates

Tank cleaning uses high pressure to jet out seawater from the nozzle of tank cleaning machine. A fraction of wash water that has been jetted out will generate liquid particulate droplets, suspending in the cabin. Most of the wash water will scour the bulkheads and their components. In the process of scouring, liquid splash and droplet splitting must take place; consequently, more liquid particulate droplets suspending in the cabin are generated.

When seawater is being jetted out from the nozzle at a high speed, seawater particles have had a quick contact and separation with the nozzle in prior. When contacting, electric double layer is formed on the interface; then during the separation, water particles take a layer of charge away from the electric double layer, and leave the other layer on the jetting lance. Therefore, the water and the nozzle respectively carries electric charge of opposite sign. This is how injection electrification works. And when seawater scouring the bulkheads, contact and separation also take place; microdroplets and bulkheads respectively carry electric charge of different sign as well. That would be the mechanism of splash electrification.

Breaking and splitting of water droplets also generate a considerable sum of electric charges. If a distilled water droplet of 8 mm breaks, it will produce 1.83×10^{-12} C of electricity on average, i.e. electric charge volume ratio is about 7.7×10^{-12} C·m^{-3}. Quantitative tests have shown that when there is no applied electric field, electricity produced by broken water droplets is approximately 3.33×10^{-6} C·m^{-3}. When there is applied electric field, the average electricity of unit volume water increases as the field strength increases. When the field strength reaches 1500 Vcm^{-1}, quantity of electrification can achieve 1.83×10^{-3} C·m^{-3}, which has increased by nearly 550 times.

Thus it can be seen that water mist formed by various liquid microdroplets inside the cabin is actually charged particulates. These particulates will certainly form a strong electrostatic field inside the cabin.

② Surge electrification of ship ballast water

Ship ballast water is usually recycled and reused to reduce the marine pollution by ships and control the amount of wastewater discharged by ships. Therefore, ballast water is inevitably mixed with oil, feculence or impurities. Experiments prove that water containing a small amount of oil has similar electrical condition as oil containing a small amount water – they both can increase the electric field strength 2~50 times. When the ship is sailing, ballast water will inevitably surge due to the shaking, and scour the bulkheads and their components; the surge itself will arouse waves which generate charged water mist. Hence ballast water can generate charged water mist then strong electrostatic field inside the cabin as well. Experiments indicate that electrostatic potential generated by ballast water of mixed cargo and passenger vessel is very close to the electrostatic potential generated when oil tankers are performing tank cleaning.

Notably, charged water mist generated by wash water or ballast water is composed of various charged suspended particles. These suspended particles have distinct properties from the continuous phase of origin liquid. First of all, Conductivity of liquid phase has a significant impact on electrification quantity; however, when they become suspended particles, these particles of discontinuous phase will acquire special electrification law where conductivity is no longer a key factor. Not only the insulated oil particles are often electrostatic when suspending, but suspended water particles, even suspended metal particles, carry a considerable quantity of electric charge. Secondly, since air is the good insulator, electric charge of suspended particles is uneasy to discharge, which makes the electric charge maintains the same span as suspended water droplets can do.

Besides, streaming electrification will take place while oil tankers are being loaded. Such electrification is similar to the case when liquid flow through pipes which has been introduced before. Therefore it is unnecessary to repeat here.

1.1.3   Electrification in Powder Industry

Powder is a special form of solid. Powder is composed of disperse solid particles; all these particles are small-sized solid dielectric. Compare with large solid materials, powder itself is characterized by dispersion and suspension. Dispersion makes the superficial area of powder many times larger than that of monoblock solid of the same material and weight; the smaller the diameter of powder particle is, the more times the superficial area will increase. Moreover, suspension of powder particles allows easy formation of smoke in the air; despite what the powder is made of, metal or insulator, suspension always makes powder particles insulated to the ground, and hence every little particle could be electrified.

The fundamental mode of powder electrification is contact electrification, which is realized by the friction and separation between powder and pipes, vessel walls and conveyer belts during fast moving, shaking or vibrating, the friction, encounter and separation with powder particles itself, and the fracture and crush of solid particles. For instance, using airflow to transport powder material, semi-manufactures and manufactures is widely accepted in plastic, chemical, food and pharmacy industries. Due to the high speed of pneumatic transportation (usually 10 m·s^{-1} on average) and high resistivity of some material, it is very easy to generate static electricity. Experiments indicate that in a simple combustible transportation system, the minimal electrification quantity of powder particles with 100 μm as their average diameter is 10^{-7}~10^{-8} C·kg^{-1}, while the maximal value can reach 10^{-4} C·kg^{-1}. Electrostatic voltage generated by powder electrification can reach thousands, even tens of thousands volts. Such voltage is extremely dangerous when combustible dust exists; little sparks of electrostatic discharge can cause violent explosions.

When powder is transported through air, powder particles collide and rub with the internal walls of pipes, and some particles even roll on the internal walls. All of these contribute to the strong electrification during pneumatic transportation. Interaction between powder particles and internal walls of pipes makes powder carry certain amount of certain electric charge, while pipe walls carry the same amount of opposite electric charge. Charge transportation between powder particles and pipe walls generates electrification current. However, it is uneasy to generate static electricity when pipes, stirrers or conveyer belts are made of the same material as powder; besides, under such condition, the polarity of powder particles is irregular as well, some of them may carry positive charge, some negative charge or none.

Similar to other natural phenomena, static electricity not just brings troubles to humans; it can also benefit mankind if appropriately utilized. Nowadays, the exploration and exploitation of electrostatic technology has become a very active technical field both abroad and at home; it has been a promising industry. For instance, electrostatic precipitation technology develops rapidly and various electrostatic dust removers can be seen all over the world; electrostatic spraying technology is popularized quickly and has been applied to shipbuilding, light industry and other industries; other technologies like electrostatic spinning and electrostatic separation are just unfolding.

1.1.4 Electrostatic Precipitation Technology

As far back as 1600, William Gillere had observed that dielectrics electrified by friction can attract the smoke that was generated by putting up candles. After about 70 years, people had successfully conducted artificial electrical discharge. In 1907, humans developed high-voltage transformer and rectifier. It was at the same time that G. Cottrell successfully manufactured electrostatic dust removers which can collect and remove the dust. Nowadays, many industrial sectors have applied electrostatic precipitation technology to deal with industrial fumes, waste gases and acid mists, which has become one of the most important technical approaches that humans possess to protect the environment and curb pollution.

The rationales behind electrostatic dust removers are as follows: electrify the dust particles by means of corona; when the particles pass through an electric field together with dust-carrying airflow, separate them from the airflow by the acting force of electric field toward charged dust particles (the Coulomb force), and then make them drifting toward dust-collecting electrons, and finally subside on the dust-collecting polar plate. The whole dust collection process can be broadly divided into four steps according to the successive physical process: the first step would be electrifying the particles; second, separate the particles with airflow; the third step would be particles subsiding on the dust-collecting polar plate; last step, transport the collected dust into the ash bucket. The electrification quantity of dust particles has a direct bearing on the efficiency of dust removers. Therefore, electrification of particles is the basis of electrostatic precipitation. Apply the high voltage to a thin wire (it is called corona wire or corona discharge electrode) then corona discharge will happen around the wire, generating large amounts of positive and negative ion pairs. Thus, there are two kinds of mechanism that both of them can electrify dust particles; one is ion collision, that is, ions that have the same polarity as corona wire are rejected, then they leave the corona wire and accelerate their motions, and collide with dust particles that electrifies the particles. The other mechanism is diffusion electrification; randomly diffused ions can electrify dust particles as well; however, it applies exclusively to tiny dust particles whose diameters are less than 0.2 μm. The process of particle electrification essentially depends on the voltage magnitude and wave form of supply voltage; after being electrified, particles also generate electric fields themselves which will first overlay with the power electric field and then in turn affect the particle electrification. Besides, with the gradual accumulation of dust layer on dust-collecting polar plate, it is inevitable to affect voltage magnitude and wave form of the effective operating-voltage of dust removers. Thus, particle electrification is a very complicated process.

Statistics show that about 60% of the dust removers are applied to power plants, 10% to steel and cement production respectively, 7% to paper-making and non-ferrous metals industry respectively, and the remaining 6% of dust removers to chemical industry, including sulfuric acid, phosphate, petroleum refining, collection of carbon black, etc..

1.1.5   Electrostatic Spraying Technology

In modern market of industrial products, almost every product has been through at least one coating process during its production, such as main parts on the shells of refrigerators and washing machines. Also, automobiles have been through at least three continuous surface spray coating process to prevent the metal plates from corrosion and make the products remain attractive in appearances. Even adjusting printing papers are coated in order to provide the surface properties required functionally. Besides, spraying technology is also applied to the production of organic solar film cells. Since the coating process is so widespread, it is of apparent great economic significance to seek for an economical, quick and safe spraying technology and craftsmanship. Precisely, electrostatic spraying technology can meet all these requirements. The most basic principle of electrostatic spraying technology is to utilize the electrostatic force, whose theoretical basis is the Rayleigh Limit. During the process of electrostatic spraying, atomized liquid is released from capillary nozzles. Due to the delta-v between nozzles and earth electrodes, electric-force gradient is formed around the liquid outlets. The gradient polarizes free charges inside the liquid, makes them move toward the applied current and be transferred onto the surface of the liquid, making liquid surface carry the same charges. Under the interaction of surface tension, gravity, potential stress of fluid surface, inertia force and viscous stress, the fluid is split into liquid droplets. After being vaporized, dried and split again, fluid droplets are finally deposited on lamina basalis. Compared with other atomization technology, electrostatic spraying has the following several advantages: it can produce liquid droplets of uniformed diameter; by carrying the same charges, Coulomb repulsion is formed, which can prevent liquid droplets from amalgamating with each other; since the diameter of spray source has a larger magnitude than diameter of liquid droplets it produced, the effect of droplets on operational process is reduced. Particularly at submicron level, electrostatic spraying technology has the advantage of producing liquid droplets/solid particles of uniformed diameter that other atomization technology can hardly match. Hence electrostatic spraying technology has following applications:

① Electrostatic spraying deposition

Deposition efficiency of charged liquid droplets can be up to 80% (Siefert 1984) under the effect of electrostatic force. Besides, by adding masks on lamina basalis, droplets can deposit precisely in the designated area (Tang et al. 2009), which can be used for film preparation.

Furthermore, with increasing attention toward environmental safety and the excellent spraying effect, electrostatic spraying technology is widely applied to insecticide spraying of crops. Damage of insecticide to the environment is the primary ecological problem worldwide. The traditional method of insecticide spraying can lead to an insecticide loss of 60~70%. Since distance can affect vaporization of liquid droplets and crop shape can affect deposition efficiency of the droplets, application of electrostatic spraying technology in insecticide spraying is the optimization of spray source and crop distance and shape.

Electrostatic spraying technology produces liquid droplet that are small in size and large in charge quantity in a high frequency, thus considered as a good method to generate homogeneous and heterogeneous structures. Generally, "deposition" is the process of generating solid film of homogeneous structure while "printing" is the process of generating heterogeneous structure according to the pre-designed pattern. Films generated by electrostatic spraying have broad applications, such as: solar batteries, fuel cells, lithium cells, micro- and nano- electronic devices, ink-jet printing and photosensitive material production. Therefore, as one of the modern printing techniques, ink-jet printing is also the industrial application of electrostatic spraying technology.

② Electrospray ionization

Electrospray ionization is the theoretical basis of using mass spectrometers to analyze macromolecules or complex molecular structures. Ions dissolved in liquid are sprayed out through charged nozzles, forming charged liquid droplets. Due to the evaporation of solvent contained in liquid droplets, diameters of the droplets reduce while charge density on droplet surfaces increases. Finally, when approaching the Rayleigh Limit, droplet surfaces start to become unstable and then release smaller charged liquid droplets; if their sizes are small enough, ions will resolve into surrounding media from droplets.

1.1.6   Electrostatic Separation Technology

Electrostatic separation technology has great potential of development. At present, it has been broadly applied to metallurgical beneficiation, grain purification, pharmaceutical separation, fiber selection and product screening, etc.

Rationale behind electrostatic separation technology is to utilize the acting force (or force moment) of high fields toward electrified objects or polarized objects to separate them. Electrostatic separation equipment vary in purpose and structure; theoretically, however, the primary step would be the same: when solid materials for separation is entering the separating area, use charged objects to electrify them with opposite charges, to electrify only one kind of particles clearly, or to make obvious differences between charges of same polarity. No matter what kind of electrification it is, its mechanism must be at least one of the followings: (1) contact electrification and triboelectrification; (2) collision electrification of ions or electrons; (3) electrification by influence.

According to different electrification mechanisms, industrial separation systems can be broadly divided into three types, as explained in detail below.

① Contact electrification and freely falling separator

Contact electrification is applicable for separating two dielectric materials. When two kinds of solid materials different in chemical constitution (such as apatite and silica) contact and separate, if the mineral fines of apatite and silica are both clean and dry, and within a temperature range of −40~800 °C, then the apatite will be positively charged while the silica negatively charged. When the mixture of these two materials are being transported to the separating area, a vertical positive-negative polar plate through conveyer belts, the two groups of particles would receive electric field force from opposite directions respectively due to their opposite polarities; therefore, their falling trajectories would deviate. Finally, separation of these two materials is achieved. It should be noted that success of the separation depends on many factors, including electric quantity of particle, applied electric field, diameter of particle, density, raw material configuration, separation temperature and electrode length. Assuming that the electric surface density of particles is 5% of atmospheric breakdown threshold and applied electric field 80%, then it can be calculated that when applying such separation, the approximate upper limit of ion radius will be 1 mm since if ion radius exceeds 1 mm, deflection effect of electric field force will be much less apparent. Inferior limit of ion radius is 20 μm; if it is less than 20 μm, the Coulomb force will combine the unseparated particles into clouds of particles, holding back the separation effect of applied electric field.

② Ion collision electrification and high voltage separation equipment

Ion collision is the most common electrification mechanism which separates good conductors from poor conductors. Currently, it is applied to separate ilmenite and rutile from silica, hematite from silica, finely divided copper wires from insulators, etc.. Capacitor discharge generates the ions for collision. The capacitor discharge electrode is made of a row of paralleled thin wires or needles; being connected to high voltage, it will perform capacitor discharge and generate positive and negative ions when electric fields around the electrode exceed the breakdown field strength of surrounding media (usually the air). Type of capacitance depends on the polarity of the electrode.

Based on such principle, raw material mixtures prepared for separation can be transported onto the grounding rotor cylinders through hoppers. When these particles pass through the electrode of strong capacitor discharge along the rotors, due to the ion collision, each material ion gets a surface electric density of σ. The quantity of σ depends on the shape of particles, raw material configuration, duration of exposure and capacitance strength, etc.. After leaving the capacitor, ions will release the surface electric charges. However, good conductors and poor conductors are different in when and how to release them. Particles with good conductivity will soon release the electric charges to grounding rotors; these charges will go into the earth and then enter the track designated by centrifugal force, gravity and air resistance. On the contrary, poor conductors will lose the electric charges slowly; then the electric field force between their surface charges and grounding rotors will adsorb the particles of poor conductors on the surface of grounding rotor to achieve the separation.

③ Electrification by influence

Electrification by influence is applicable to separate good conductors from good insulators. In certain conditions, it is also applicable to separate at least two kinds of semi-conductors of very different conductivity. Existence of high voltage electrode will generate a strong electric field to electrify the grounding rotors. Conductor particles in admixtures fallen from V-shaped hoppers will quickly catch electric charges of high quantity and then be adsorbed toward the high voltage electrode. Non-conducting particles of hybrid particles are only slightly electrified; therefore, the separation can be achieved.

1.2   Electrification of Particulate in Natural Phenomenon

Although surrounded by the atmosphere, humans may not be able to feel the existence of atmospheric electrostatic field. Still, the atmospheric electrostatic field does exist, and influence our lives in its own peculiar way. A large number of observations have shown that even on sunny days, the atmosphere is filled with electrostatic fields. Being in such electrostatic fields, human bodies (conductor) also generate current flows; however, the magnitude of such current is only 1pA, which is considerably lower than the critical value of human perception. Test results concluded by many scientists have shown that the magnitude of atmospheric electrostatic field strength near the terrain surface is about 100 V/m, with the direction of top down, being perpendicular to the ground. The higher the ground is, the weaker the field strength becomes.

There are a variety of transformations from gas to solid in nature. Electric fields are often generated along with these natural processes, such as thunderstorm, snowstorm, hail, sandstorm and dust-whirl. People can see the outward manifestations of atmospheric electrostatic field from these natural phenomena, and also get a taste of the enormous power of it.

1.2.1   Electrification in Wind-Blown Sand

In natural granular flows, such as wind-blown sand flows, snowstorms, volcanic plumes etc., the existence of electrostatic charges has been widely acknowledged. Measurements of terrestrial dust electrification at the surface and aloft indicate that particle charging in dust storms is a common, and, perhaps, even a universal phenomenon to a greater or lesser extent (Harrison et al. 2016). Gill has observed that there appears strong electric field and electric spark in the sandstorm transit zone; meanwhile, radio signals are interfered. When measuring the electric field of dust-whirl, Farrell and others have found that the electric field strength reaches 4 kV·m^{-1} within 50 m, which is much higher than what people have estimated (Farrell 2004).

Since the beginning of the last century, numerous scholars have attempted to understand the mystery of the electrification of wind-blown sand utilizing a variety of methods and aspects, including field and wind tunnel measurements, theoretical analysis on the particle charging mechanism, and quantitative predictions derived from theoretical models. A general consensus has been reached that the E-field of wind-blown sand is produced by moving sand particles with opposite electric polarity, where the polarity is somehow related to the size of the sand particles.

Movement of sands inside the dust-devil shows an obvious phenomenon of stratification: the small solid particles are distributed around the periphery of dust-whirl; the medium solid particles are in the inner layer of wind field. Small and medium solid particles first drop down to the ground from periphery and inner layer of dust-whirl, then get carried to the bottom center of dust-whirl by surficial airflow. Since the updraft is unable to carry large solid particles into the air, they always hover over the bottom of wind field. Due to the low volume fraction and the stratification, sand grains of dust-whirl seem unlikely to encounter in mid-air; however, during the circulating process, medium and small solid particles will re-enter into the bottom confluence area, which will considerably increase the probability of colliding with large solid particles. Besides, all sand grains will collide and rub against the ground; thus sand grains and the ground is obviously charged.

1.2.2 Electrification in Thunderstorm

Thunderstorm is a kind of regional storm formed by cumulonimbus. The happening of thunderstorm means that the strongest atmospheric convection conducts the strongest discharge and release enormous energy. Because thunder is produced by lightning, defining a thunderstorm as a cloud that produces thunder means that it is also a cloud that is electrified enough to produce lightning.

Most scholars who study on atmospheric static electricity consider that, compared with its surrounding cloudless air, cloud is more insulated. However, when it comes to the electrification mechanism of thundercloud, they cannot reach an agreement. Still, it is recognized that charge separation process must exist as thundercloud is macroscopically charged. Laboratory experiments and theoretical studies have indicated that such charge separation may be the result of particle encounter, selective ion attachment (ion trapping), micro-droplets melting or micro-droplets condensation. As for which mechanism plays the leading role or how important each mechanism can be relatively, there is no final conclusion yet. Among which, the most acceptable one would be the Theory of Ion Trapping proposed by C. T. R. Wilson. The theory states that there are various sizes of droplets inside the cloud; larger droplets drop faster than smaller ones in the air; once there generates small amount of electric charge in the cloud, the lower surface of droplets will induce positive charge while the top surface will induce negative charge. When droplets are dropping, they give the positive charge to the air they fall through. Therefore, droplets usually trap more anions than cations in the air. Anions left by larger droplets will be gathered by smaller droplets first, then they will be brought to the head of cloud by airflow, and finally an electrostatic field is formed. Another theory states that the reason why cloud is electrified is that water will generate electric charge when it is freezing. Laboratory experiments have showed that when aqueous solution is freezing, there will be a large potential difference between water and ice. Water gets positive charge while ice gets negative charge. Updraft in the ice zone takes away the water that is above ice particles and leads to the separation of electric charge. Small droplets are brought to the top by updraft, which makes the head of cloud positively charged while larger ice particles drop to a lower altitude and are negatively charged.

Due to the complexity of thundercloud structure and electrification mechanism, and the difficulty of carefully observation into clouds, many theories for the formation mechanism of charge on cloud particles can only be derived from the laboratory experiments and field observations. There are more than ten theories including the two aforementioned two mechanisms, but all are hypotheses, and no any individual can successfully explain the observation results. It's now commonly believed that the thunderstorm electrification results from the synthetic action of multiple mechanisms, and the dominant mechanism varies with the developing stages of clouds.

1.3 Influence of Particulate Electrification to Multiphase Flows

Safety hazards and product hazard caused by static electricity can appear as a variety of patterns such as electrostatic Coulomb force, electrostatic discharge and static induction, among which electrostatic discharge is the most severe and common hazard.

1.3.1 Influence of Coulomb Force

Electrostatic charges accumulated on objects will generate electric fields in space around them. Electrostatic Coulomb force will affect objects in the electric field.

When the object in electric field is electrified, the quantity and direction of the electrostatic Coulomb force is described by the Coulomb Law. When the object in the electric field is uncharged, if it is a conductor, static induction will take place; if it is a insulator, polarization will take place. In either case, the object will be acted upon by a certain force, and rotate under the effect of that force to make its length direction in accordance with field direction.

If the electric field is uniform, the Coulomb force acting upon induced charges or polarized charges will not make the object shift its position. However, if the electric field is not uniform, the Coulomb force acting upon induced charges or polarized charges will make the object move toward the direction of the electric field. In other words, in non-uniform electric fields, uncharged objects will be adsorbed to electrified bodies due to the strong electric fields around the electrified bodies. The higher the inhomogeneity of electric field is, the greater the adsorption is. Generally, the magnitude of electrostatic force generated by objects is few Newton per square meter. Although only ten thousandth of magnet force, it can perform obvious adsorption toward light objects including hairs, paper scraps, dust and fiber. It is the adsorption of Coulomb force that brings various hazards to different industries, different production environment and conditions, and different products.

During the production, repulsive effect of charges with the same electrical sign and sucking action of opposite charges can also cause production hazard. For instance, in textile industry, when reeling off raw silk from cocoons, electrostatic force will make raw silk flutter and tangle; during the weaving process, friction between rubber rollers and yarns can generate electrostatic force muddling, hanging and tangling the yarns, which will lower the carding power of needle gears and eventually affect product quality and production efficiency. In the powder processing industry, electrostatic force can lessen or clog the meshes, making pneumatic pipelines less smooth. In printing industry and plastic film packaging production, suction force or repulsive force generated by static electricity will negatively affect the normal processes of paper separation, stacking, and it will also make plastic film unable to be packaged and printed as normal or even generate "electrostatic ink spots", which will bring troubles to automatic production.

1.3.2 Influence of Electrostatic Discharge

The accumulation of electrostatic charge on the objects, from the perspective of energy, manifests as the storage of electrostatic energy and electric fields will be generated around the object. When the electric field strength (the electric field strength of two charged objects is determined by the disparity and distance of the objects) between the potential of the charged object and the lower potential of the nearby object exceeds disruptive strength of the media between these two objects, electrostatic discharge will be produced for leakage power. Especially in the case of non-uniform electric field, high electric field strength will be formed in some areas, which makes discharge more easily. Upon the formation of electrostatic discharge, the stored electrostatic energy will be released rapidly. At the same time, high current density will appear instantly, accompanied by sound, light, heat and electromagnetic radiation, resulting into different hazards. Thus, electrostatic discharge works as the main mode of static electricity hazards. Typical electrostatic discharge hazards usually include the following aspects:

1.3.2.1 Causing Fires and Explosive Accidents

Electrostatic discharge, serving as the ignition source, will not cause fires and explosion unless the following three conditions are met simultaneously: sparks are produced during electrostatic discharge; flammable gases or the mixture of combustible dust and air exist in the spark gap of electrostatic discharge and are within the explosion concentration limit; and electrostatic discharge energy is greater than or equal to the minimum ignition energy of the explosive mixture. As long as the above three conditions are simultaneously satisfied, then it's possible to cause combustion and explosion; whether the accident will happen is just a matter of probability.

Probability of explosive accidents caused by electrostatic discharge depends on discharge energy. Usually, electrostatic discharge can be divided into spark discharge, brush discharge and surface discharge (also known as discharge along surface or propagating brush discharges) as well as corona discharge, among which spark discharge marks the most dangerous one, followed by surface discharge and brush discharges. With large discharge energy, the above discharges are less likely to form an ignition source.

In the case of electrostatic discharge caused by electrified conductors like metals, nearly all of the stored electrostatic energy will be released usually. During this process, when the accumulated energy is greater than the minimum ignition energy of combustible material, such an electrostatic discharge form is more likely to be an ignition source to trigger fire and explosion. When static effect happens in insulators such as plastic, chemical fiber, rubber-like, such insulators, with a low conductivity, can only release some parts of its stored static electricity. Such insulators, having a relatively high limit of electrostatic potential, are believed to have a low risk, especially when their potential is less than 5 kV. However, with abundant stored electrostatic charge and large surface charge density, the above insulators are possible to generate a surface discharge, which makes them more dangerous (Table 1.1).

In addition, the ignition capability of discharge is also relative to polarity. With the same potential, the ignition capability to combustible material caused by discharge when the surface of liquid or solid contains negative charge is one order of magnitude higher than that containing positive charge. Moreover, when the temperature of combustible material is higher than normal temperature, oxygen content in the local environment (or other oxidant gas content) is higher than that in the normal air, or the pressure of explosive gases is higher than the atmospheric pressure, it is also more likely to cause combustion.

1.3.2.2 Causing Electric Shock on Human Body

When human body approaches electrostatic object or human body with static electricity approaches grounding conductors, as long as the electric field strength calculated by the ratio between the potential difference and the distance exceeds the insulating strength of the distance, electrostatic discharge will take place. Instantly, a shocking current will be formed and pass through the human body, causing an electric shock. The degree of electric shock is relative to the stored electrostatic energy. The greater the energy is, the greater the earth capacity or voltage of charged body will be, and hence, the more severe the electric shock will be.

Although electric shock caused by static electricity in the normal course of the production process will not kill people usually, it is possible to lead to fingertip injury, finger numbness, or other functional damages causing phobias. More

Table 1.1 Features of various types of discharge and relative ignition capability

Type of discharge	Occurrence conditions	Features & ignition capability
Corona discharge	It happens easily when electrodes are far apart from each other, or when the tip or the ledge of the object surface has a strong electric field.	Accompanied by sound and light sometimes, discharge channel will not be formed when gaseous medium conducts local ionization around the tip of objects. Single pulse corona discharge energy of induction corona is less than 20 μJ; On the contrary, single pulse corona discharge energy of active corona is a few times larger. Under such circumstance, the ignition capability is quite small.
Brush discharge	It happens easily between electrostatic non-conductors with high potential and conductors.	Accompanied by sound and light, discharge channel forms bifurcations around surface of electrostatic non-conductor. During this process, less energy is released in a particular unit of space. Usually discharge energy released every single time does not exceed 4 mJ. Under such circumstance, the ignition capability is at intermediate level.
Spark discharge	It mainly happens between charged metal conductors in short distance.	Accompanied by sound and light, the discharge channel usually does not form bifurcations. With a clear discharge focus on electrodes and more concentrated discharge energy, the ignition capability is quite strong.
Propagating brush discharges	It only happens in the case of high-speed discharge. When the thickness of electrostatic non-conductors is less than 8 mm and the surface charge density is greater than or equal to 2.7×10^{-4} $C \cdot m^{-2}$, it happens easily.	Accompanied by sound and light during discharge, a large number of electrostatic charges within a certain range of electrostatic non-conductors will be released. With great discharge energy, the ignition capability is quite strong.

importantly, it may give rise to secondary damages such as falling from high altitude or falling down.

1.3.2.3 Causing Interference to the Normal Operation of Electronic Equipment

In the process of electrostatic discharge, frequency band ranging from hundreds of kilohertz to dozens of megahertz will be produced. Besides, broadband electromagnetic interference (hereinafter referred to as "EMI") with dozens of milli-volts will also be generated. Such interference may be coupled to low level digital circuits of computers and other electronic equipment through a variety of ways, which can upset circuits and lead to malfunctions. Clutter caused by electrostatic discharge can directly go into devices or receiver circuit through capacitive coupling, inductive coupling or relevant signal channels, which will not only cause circuit malfunctions, but also intermittent or interfering failure, information loss, or temporary damage to certain functions. However, it may cause no obvious damage to the hardware. Upon termination of electrostatic discharge and interference, devices and equipment are likely to return to normal. Devices and equipment, if new work signal is inputted again, can still start and continue working. For example, discharge performed by large-scale sandstorms will cause spark discharge on high-voltage wires, tripping of power transmission network and interference to monitoring electronic equipment.

1.3.3 Influence of Electrostatic Induction

Around the electrostatic charged body, within the scope of electric field force, induced charges will be produced on isolated (namely insulated from the earth) conductors or semiconductor surface in this area. Among which the surface close to the charged body will generate charges opposite to the sign of the charged body, while the other surface will generate charges that have the same sign with the charged body. As the entire conductor is insulated from the ground and charges do not leak, the positive and negative charges carried by the conductor can maintain a state of equilibrium due to the electric field of the charged body, while the total quantity of electric charges is zero. However, the complete separation of positive and negative charges on the surface of the conductor fully enables it to share the property of electrostatic charge. Apparently, the potential amplitude of this conductor depends on the electric field strength of the original charged body.

Electrostatic induction works as a way to electrify objects. Thus, induced charged body can not only produce Coulomb force adsorption, but also form electrostatic discharge with its adjacent objects, and cause various harms of these two types of models.

References

Farrell, W.M. 2004. Electric and magnetic signatures of dust devils from the 2000–2001 MATADOR desert tests. *Journal Of Geophysical Research* 109: E03004.

Harrison, R.G., E. Barth, F. Esposito, J. Merrison, F. Montmessin, K.L. Aplin, C. Borlina, J.J. Berthelier, G. Déprez, W.M. Farrell, I.M.P. Houghton, N.O. Renno, K.A. Nicoll, S.N. Tripathi, and M. Zimmerman. 2016. Applications of electrified dust and dust devil electrodynamics to martian atmospheric electricity. *Space Science Reviews* 203 (1): 299–345.

Siefert, W. 1984. Corona spray pyrolysis: A new coating technique with an extremely enhanced deposition efficiency. *Thin Solid Films* 120 (4): 267–274.

Tang, J., E. Verrelli, and D. Tsoukalas. 2009. Assembly of charged nanoparticles using self-electrodynamic focusing. *Nanotechnology* 20 (36): 365605.

Chapter 2
Properties of Particulate in Multiphase Flows

2.1 Particulate Forms

The substance which has the same composition, homogeneous physical and chemical properties is called a "phase". In other words, a "phase" is a set of a single substance (Ruspini et al. 2014). Many particulate-entrained fluid flows in nature and industry are multiphase systems. Gas-droplet, gas-particle and liquid-particle flows in which the particles and droplets constitute the dispersed phase can be called dispersed phase flows.

Wind-blown sand is a typical gas-solid two-phase flow, which is composed of wind (fluid phase) and sand (solid particle). Electrospray is usually a gas-liquid two-phase flow, which is composed of gas (fluid phase) and droplet (liquid particulate). Note that particulates occurring in particulate-entrained fluid flows have two basic forms – solid phase (particle) and/or liquid phase (droplet).

2.2 Particulate – Fluid Interaction

Particulate – fluid interaction refers to the exchange of properties between phases and is responsible for coupling in dispersed phase flows (Crowe et al. 2011).

1. Mass coupling

Mass coupling can occur through a variety of mechanisms such as evaporation, condensation or chemical reaction. The mass transfer from a slurry droplet represents an important technological problem. For example, powdered milk to be dried consists of water and solids. These slurries are atomized and sprayed into a hot gas stream where the water is driven off and the dried products are collected. The droplet material can be thought of as a porous medium formed by the solids. As the

© Springer Nature Singapore Pte Ltd. 2017
Z. Gu, W. Wei, *Electrification of Particulates in Industrial and Natural Multiphase flows*,
DOI 10.1007/978-981-10-3026-0_2

drying proceeds, the size of the droplet may not change appreciably (it may actually increase slightly in diameter), but the mass decreases as the moisture is removed.

2. Linear momentum coupling

Linear momentum coupling between phases occurs as the result of mass transfer and interphase drag and lift.

A rigorous derivation of the equation of motion for small particles in non-uniform, unsteady flows at low Reynolds numbers was (Maxey and Riley 1983):

$$
\begin{aligned}
m\frac{dv_i}{dt} &= mg_i + V_d\left(-\frac{\partial p}{\partial x_i} + \frac{\partial \tau_{ij}}{\partial x_j}\right) + 3\pi\mu_c D\left[(u_i - v_i) + \frac{D^2}{24}\nabla^2 u_i\right] \\
&+ \frac{1}{2}\rho_c V_d\frac{d}{dt}\left[(u_i - v_i) + \frac{D^2}{40}\nabla^2 u_i\right] \\
&+ \frac{3}{2}\pi\mu_c D^2 \int_0^t \left[\frac{d/d\tau\left(u_i - v_i + D^2/24 \times \nabla^2 u_i\right)}{\pi\nu_c(t - \tau)^{\frac{1}{2}}}\right] d\tau
\end{aligned}
\tag{2.1}
$$

where m is the particle mass. The first term on the right is the body force due to gravity. Any additional body forces would have to be included here. The terms contributing to the fluid dynamic force on the particle are (1) the pressure and shear stress due to the undisturbed flow, (2) the steady state drag, (3) the virtual or apparent mass term and finally (4) the Basset or history term. The last two terms are operative only in unsteady flows. Each of these terms will be addressed individually and rewritten for conditions outside the realm for which Maxey and Riley's equation is valid.

When there is no acceleration of the relative velocity between the particle and the conveying fluid, the steady-state drag force in Maxey and Riley's equation can be re-expressed as

$$
F_{ss,i} = \frac{1}{2}\rho_c C_D A|u_i - v_i|(u_i - v_i)
\tag{2.2}
$$

where C_D is the drag coefficient, A is the representative area of the particulate and u_i and v_i are the velocities of the continuous phase and the dispersed phase, respectively. Typically the area is the projected area of the particulate in the direction of the relative velocity. In general, the drag coefficient will depend on the particle shape and orientation with respect to the flow as well as on the flow parameters such as Reynolds number, Mach number, turbulence level and so on.

Lift forces on a particle are due to particle rotation. This rotation may be caused by a velocity gradient or may be imposed from some other source such as particle contact and rebound from a surface.

The Saffman lift force is due to the pressure distribution developed on a particle in a velocity gradient. The higher velocity on the top of the particle gives rise to a low pressure, and the high pressure on the low velocity side gives rise to a lift force. The magnitude of force for low Reynolds numbers to be (Saffman 1965):

$$F_{saff} = 1.61\mu_c D |u_i - v_i| \sqrt{Re_G} \tag{2.3}$$

where Re_G is the shear Reynolds number defined as

$$Re_G = \frac{D^2}{\nu_c} \frac{du}{dy} \tag{2.4}$$

This can be thought of as the Reynolds number based on the velocity difference between the bottom and top of the particle.

The Magnus force is the lift developed due to rotation of the particle. The lift is caused by a pressure differential between both sides of the particle resulting from the velocity differential due to rotation. The rotation may be caused by sources other than the velocity gradient. If the rotation vector is normal to the relative velocity vector then the Magnus lift force is

$$F_{Mag} = \frac{1}{2}\rho_c C_{LR} A |v - u|(v - u) \tag{2.5}$$

where A is the projected area of the particle and C_{LR} is the lift coefficient due to rotation.

2.3 Particulate – Particulate Interaction

Dispersed phase flows consist of a large number of solid particles. These particles may have a size distribution ranging from a few micrometers to centimeters. The dynamics of the granular material, or the solid phase in multiphase flow, is governed by the Newton's second law of motion for the center of mass of each particle and by the Euler's second law of motion for the angular momentum change.

Particle–particle interaction controls the motion of particles in dense particle flows. Also particle-wall interaction is important in dense flows as well as wall-dominated dilute flows. Particle-particle collision is unimportant in dilute gas-particle flows. As the particle concentration becomes higher, particles collide with each other and the loss of particle kinetic energy due to inter-particle collision cannot be neglected. With respect to particle–particle interactions in multiphase flow dynamics, two phenomena are identified: collision and contact. From the viewpoint of physics, collision and contact do not differ significantly. Collision is merely contact with short time duration; however, the modeling approach is different for each. For collision or contact, two models are normally used, the hard sphere model and the soft sphere model.

The hard sphere model is easy to use but applicable only to binary collisions. The relation between the pre- and post-collision velocities is given explicitly using the coefficient of restitution and friction coefficient. Fortunately, as long as the particulate phase is dispersed, it is sufficient to consider only simple binary collisions and not multiple collisions.

The soft sphere model is modeled by using mechanical elements such as a spring and a dash-pot. The soft sphere model is called as DEM (Discrete Element Method or Distinct Element Method). The technical term "DEM" is more popular than "soft sphere model." In the soft sphere model, the whole process of collision or contact is solved by numerical integration of the equations of motion. The computation time is much longer in the soft sphere model than in the hard sphere model, but the applicability for the soft sphere model is wider than the hard sphere model, especially in the field of particulate electrification by using numerical modeling methods.

The soft sphere model was developed for the purpose of simplification (Cundall and Strack 1979). First, it is assumed that the influence of the particle of interest is limited to only neighboring particle(s) which is (are) in direct contact with that particle. Next, the deformation is replaced with the overlap of two particles. That is, instead of dealing with the deformation, the particles contacting each other under the influence of external forces are made to overlap. The more the overlap distance, the larger the repulsive force. This repulsive force can be expressed by equating the overlap distance with compression of a spring. In general, deformation of a body accompanies energy loss. This is the reason why the coefficient of restitution is less than one. The energy loss depends on the deformation speed. A dash pot or a viscous damper is a suitable mechanical element to model the energy loss.

Figure 2.1 shows two particles i and j with position vectors $\vec{x}_i$, $\vec{x}_j$, and radii R_i and R_j, which are in physical contact. Normal overlap of two particles is defined by:

$$\delta_n = R_i + R_j - \left| \vec{x}_j - \vec{x}_i \right| \tag{2.6}$$

If $\delta_n > 0$, there is a physical contact between particles. The collision force between the two particles is comprised of normal and tangential collision forces represented as f_{ij}^n and f_{ij}^t, respectively

$$\vec{f}_{ij}^c = \vec{f}_{ij}^n + \vec{f}_{ij}^t \tag{2.7}$$

The normal vector is a vector points from the center of particle i to the center of particle j and is defined as:

$$\vec{n}_{ij} = \frac{\vec{x}_j - \vec{x}_i}{\left| \vec{x}_j - \vec{x}_i \right|} \tag{2.8}$$

The relative velocity between colliding particles at the contact point and its normal and tangential components are defined as:

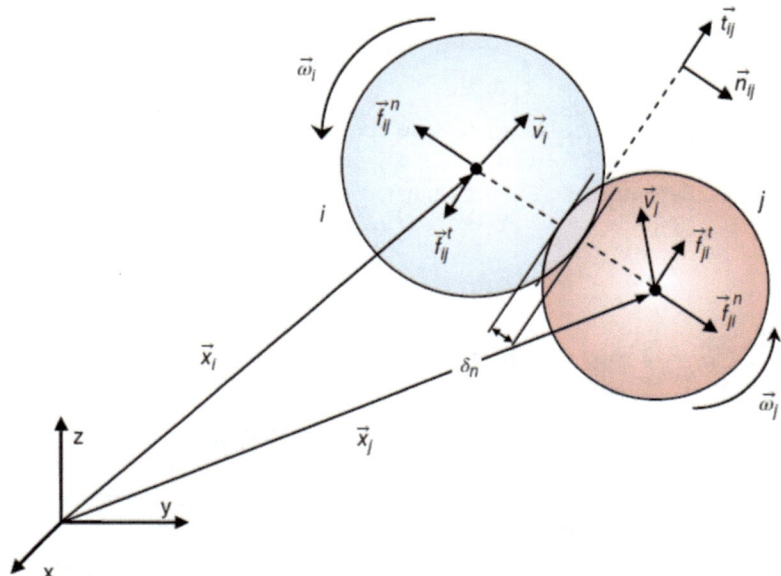

Fig. 2.1 Schematic illustration of colliding particles i and j and forces acting on particles as a result of collision (Norouzi et al. 2016)

$$\vec{v}_{ij} = \vec{v}_i - \vec{v}_j + \left(R_i \vec{\omega}_i + R_j \vec{\omega}_j \right) \times \vec{n}_{ij} \tag{2.9}$$

When there is a physical contact between any two particles, force-displacement laws are used to calculate collision properties as a function of normal and tangential overlaps, physical properties, and collision history of the colliding particles. A comprehensive model that accurately computes the collision forces needs complicated contact mechanics and its implementation is really hard and time consuming. Since the common granular or multiphase flow consists of more than thousands of particles, there are numerous collisions for which collision forces should be evaluated in each time step. This is not very practical for our purposes. Therefore, simplified models, called force-displacement laws, are utilized to reduce the computational effort and keeping the accuracy at a level acceptable for engineering applications.

There are many force-displacement laws developed and applied to DEM, for example, linear viscoelastic, nonlinear viscoelastic, and elastic perfectly plastic models. More details on force-displacement models and their characteristics can be found elsewhere (Di Renzo and Di Maio 2004; Kruggel-Emden et al. 2007; Thornton et al. 2011).

2.4 Basic Theory of Electrostatics Induced by Particulate Electrification

Electrostatics is the inheritance and development of many other related disciplines. Therefore, many physical quantities showing physical properties of static electricity are formed, developed and conceptualized on the basis of electricity and other disciplines.

2.4.1 Charge and Coulomb's Law

Charges and their interaction are the basis of the formation and existence of all electromagnetic phenomena. The movement of charges can produce currents and generate a magnetic field around the charges. The amount of charges is called quantity of electric charge. In order to study the charge distribution on the object, volume charge density, surface charge density and charge-mass ratio are introduced to describe electrification of particles.

1. Volume charge density

Volume charge density is defined as the quantity of electric charge per unit volume, reflecting the charged situation of the object volume.

$$\rho = \frac{Q}{V} \tag{2.10}$$

In the formula,

ρ volume charge density, $C \cdot m^{-3}$;
Q quantity of electric charge, C;
V volume of the charged body, m^3

2. Surface charge density

Surface charge density is defined as the quantity of electric charge per unit surface area, reflecting the charged situation of the object surface.

$$\sigma = \frac{Q}{S} \tag{2.11}$$

In the formula,

ρ volume charge density, $C \cdot m^{-2}$;
Q quantity of electric charge, C;
V surface area of the charged body, m^3

3. Charge-mass ratio

Charge-mass ratio is the ration between the quantity of electric charge and the mass of charged body, namely the quantity of electric charge of objects per unit mass. It is usually used to show the charged condition of objects.

$$\rho_m = \frac{Q}{m} \tag{2.12}$$

In the formula,

ρ_m Charge-to-mass ratio, $C \cdot kg^{-1}$;
Q quantity of electric charge, C;
m mass of charged particles, kg.

When gradually adding negative charges to a particle with positive charges, positive charges of this particle will gradually reduce firstly and then completely disappear. Only after a complete loss of positive charge, the particle begins to show its negatively charged property. On the contrary, a particle with negative charges can only carry positive charges when its negative charges begin to reduce gradually and finally completely disappear. Thus, we know that heterocharge can neutralize with each other. In the process of triboelectrification, two uncharged particles, after friction, are both charged. Besides, one particle is always positively charged and the other negatively charged. In the process of electrostatic induction, the positively induced charges and negatively induced charges are always generated simultaneously, and their quantities are always equal. In uncharged objects, there is always the same quantity of positive and negative charges. With the equal quantity of positive and negative charges, the object does not show electrical property. To make the object positively or negatively charged, to be specific, is to make the quantity of positive charges or negative charges carried by the object exceed their quantities in neutralization. Such neutralized quantity of charges can be called excess charge or net charge.

Charge can only be transferred from one object to another, or transferred from one part of the object to another part of the object. However, charge can neither be created nor vanished, which is called the law of conservation of charge. In an isolated system, no matter what kind of physical processes occur, the algebraic sum of quantity of electric charge in the system remains unchanged.

All matter is composed of atoms. Normally, the number of electrons outside the nucleus is equal to the number of protons inside the nucleus, so the atom is electrically neutral. Thus, the entire macroscopic object is also electrically neutral. If the atoms or molecules, due to the external effects, lose one or several electrons, the atoms or molecules will become positive ions. On the contrary, if the atoms or molecules, due to the external effects, gain one or several electrons, the atoms or molecules will become negative ions. Under certain external effects, macroscopic objects (or one part thereof) gain or lose a certain number of electrons, causing the total number of electrons are no longer equal to that of protons, the object will show its electrical property. Then, the object is considered electrostatically electrified or to be in an electriferous condition.

Experiments have shown that charges always appear in an integer multiple of a basic unit. Such a property is called the quantization of charge. The basic unit of the charge is the absolute value of the quantity of electric charge carried by an electron, which is usually represented by e.

Coulomb's law can be used to describe the interaction force between charged bodies. The magnitude of interaction force between two point charges is proportional to the product of their electric quantity, and is inversely proportional to the square of their distance. The force is along the straight line joining them. If the two charges have the same sign, the electrostatic force between them is repulsive; if they have different signs, the force between them is attractive.

$$F_{12} = \frac{q_1 q_2}{4\pi\varepsilon_0 r^2} \vec{r}_{21} \tag{2.13}$$

In the formula,

F_{12} the acting force of one point charge on another, N;
q_1, q_2 the quantity of electric charge of these two point charges, C;
$\vec{r}_{21}$ the unit vector of one point charge on another;
ε_0 permittivity of vacuum, $\varepsilon_0 = 8.85 \times 10^{-12}$ F·m^{-1};

If two point charges are in a homogeneous medium, then the permittivity (hereinafter referred to as "ε") will be employed to replace the permittivity of vacuum (hereinafter referred to as "ε_0"). The interaction force between charges is also called the Coulomb forces or electrostatic forces, among which the Coulomb forces are vectors, satisfying the superposition principle.

2.4.2   Electric Field and Gauss Theorem

1. Electric field

All charged objects create electric fields that extend outwards into the space that surrounds them. The electric charges alter that space, causing any other charged object that enters the space to be affected by this field. The coverage that the Coulomb force of electric charges affects is called an electric field where an electric charge interacts with another.

Electrostatic fields are electric fields triggered by stationary charges and currents. Temporally, both electrostatic fields and their intensity remained constant.

The electric field intensity quantitatively describes the vector quantity of electric field at a point around each electric charge. It is equal to the force received by per electric charge at a point, and its direction is that of received force. The electric field intensity can be calculated using the formula shown below:

$$\vec{E} = \frac{\vec{F}}{q_0} \tag{2.14}$$

In the fornula,

$\vec{E}$ is electric field intensity (unit: $V \cdot m^{-1}$).

$\vec{F}$ is Coulomb force (unit: N).

q_0 is the quantity of per positive charge (unit: C).

The superposition principle of electric field intensity defines that the total electric field intensity at a point equals the vector sum of electric fields created by multiple electric charges at this point. This principle can be used to calculate the electric field intensity of charged bodies.

2. Gauss theorem

An electrostatic field is a vector field whose property is indicated by the flux extending along a closed surface and circular vector extending along the closing route.

The surface integral of electric field intensity on a given surface in an electric field is the electric flux and is represented by ϕ_e. Gauss theorem defines that, in a vacuum, the flux extending along a closed surface equals the ratio of the algebraic sum of enclosed charges against vacuum permittivity. The ratio can be calculated using the following formula:

$$\phi_e = \oint_S \vec{E} \cdot d\vec{S} = \frac{\sum q}{\varepsilon_0} \tag{2.15}$$

If q is a positive charge and $\phi_e > 0$, there are electric power lines generated from q penetrating out of the closed surface. Therefore, q is a source of an electrostatic field. If q is a negative charge, there are electric power lines penetrating into the closed surface and are terminated at q. It can be inferred that an electrostatic field has a source. Gauss theorem elaborates the relationship between the electric field and the source.

To find out the relationship between a point in an electric field and the neighboring flux, diminish the closed surface so that ΔV that contain this point is close zero. Then, calculate the limit value using $\lim\limits_{\Delta V \to 0} \frac{\oint_S \vec{E} \cdot d\vec{S}}{\Delta V}$. This limit value is the divergence of field intensity vector $\vec{E}$ and is represented by $\mathrm{div}\,\vec{E}$ or $\nabla \cdot \vec{E}$ which can be calculated by the following formula:

$$\mathrm{div}\,\vec{E} = \lim_{\Delta V \to 0} \frac{\oint\limits_{S} \vec{E} \cdot d\vec{S}}{\Delta V} \tag{2.16}$$

It can be inferred that the electric field intensity at a point in an electric field is the limit value of electric flux density of this point.

Use Gauss theorem to this minimized closed surface to calculate the divergence $\vec{E}$ as follows:

$$\nabla \cdot \vec{E} = \lim_{\Delta V \to 0} \frac{\oint\limits_{S} \vec{E} \cdot d\vec{S}}{\Delta V} = \lim_{\Delta V \to 0} \frac{\Delta q/\varepsilon_0}{\Delta V} = \frac{\rho}{\varepsilon_0} \tag{2.17}$$

Formula (2.17) points out that the divergence of electric field intensity $\vec{E}$ at a point equals to the ratio of charge density ρ at the point against vacuum permittivity ε_0.

For uniform dielectric, the differential form of Gauss theorem is:

$$\nabla \cdot \vec{E} = \frac{\rho}{\varepsilon} \tag{2.18}$$

Formula (2.18) points out that if $\rho > 0$ at a point in an the electric field, which are positive charges at this point, meaning $\nabla \cdot \vec{E} > 0$, indicating an $\vec{E}$ flux is generated; if $\rho < 0$ at a point, which are negative charges, meaning $\nabla \cdot \vec{E} < 0$, indicating that $\vec{E}$ flux terminates at this point; if $\rho = 0$ at a point, which are no charge, meaning $\nabla \cdot \vec{E} = 0$, indicating that $\vec{E}$ flux is continuous and the electric field lines pass through the point. The positive charge generates $\vec{E}$ flux and is a positive source, while the negative charge absorbs $\vec{E}$ flux and is a negative source. Divergence formula (2.18) of $\vec{E}$ reflects the relationship between the flux and its sources at a point and is known as the differential form of Gauss theorem for electrostatic fields.

2.4.3   Electric Potential and Basic Electrostatic Field Equations

In an electrostatic field, electrostatic force drives charges to move, and the electric field works on the charges. This is similar to the effect of a gravitational field on an object, by generating electrostatic and electric potential. Electric potential energy is a relative potential energy amount of a charge at a point in an electric field. In this case, there must be a zero reference point. Generally, if a charge is distributed in a limited area, the electrostatic potential energy of charge q_0 at infinity distance is zero, that is, $W = 0$. Then, the electrostatic potential energy of charge q_0 at A point in an electric field can be obtained with a value of work $A_{a\infty}$ made by charge q_0

moving from A point to infinity distance. Based on the relationship of the work and force, $A_{a\infty}$ can be calculated through the following formula:

$$W_a = A_{a\infty} = q_0 \int_a^\infty \vec{E} \cdot d\vec{l} \tag{2.19}$$

If charge q_0 moves from A point to B point, the work A_{ab} made by the electric field can be calculated through the following formula:

$$A_{ab} = W_a - W_b = q_0 \int_a^b \vec{E} \cdot d\vec{l} \tag{2.20}$$

Formula (2.20) shows that the work done by the electric field force when a charge moves in an electrostatic field is related only to the quantity of the charge q_0 and the start and end points of the movement, despite of the route. If the charge returns to the start point along the closing route, the work made by the electric field is zero.

Electrostatic potential energy W_a cannot directly describe the property of A point in the electric field, because the W_a is related to q_0. Therefore, the ratio W_a/q_0 is used to indicate the property of A point in an electrostatic field. The property of A point is determined only by the property of the electric field and the location of A point. The following formula shows the mathematical relationship.

$$V_a = \frac{W_a}{q_0} = \int_a^\infty \vec{E} \cdot d\vec{l} \tag{2.21}$$

V_a is the potential of A point. If q_0 is a positive charge, $V_a = W_a$, that is, the potential value of a point equals to the electric potential energy per positive charge at this point and equals to the work made by this point when moving along a route to infinity distance.

The potential difference of V_A and V_B can be calculated using the following formula:

$$V_A - V_B = \int_A^B E \cos\theta dl \tag{2.22}$$

If A point is close to B point with a distance of Δl, the potential increment of A and B is ΔV, V_A minus V_B equals $-\Delta V$ and El is at the direction component of electric field E, then:

$$-\Delta V = El\Delta l \tag{2.23}$$

If $\triangle l \rightarrow 0$, the formula is:

$$El = -\frac{\partial V}{\partial l} \tag{2.24}$$

If l denotes the direction of x, y, and z axis, the component of electric field intensity E in three directions can be obtained as follows:

$$E_x = -\frac{\partial V}{\partial x}; E_y = -\frac{\partial V}{\partial y}; E_z = -\frac{\partial V}{\partial z} \tag{2.25}$$

Electric field intensity E can be obtained through:

$$\vec{E} = E_x\vec{i} + E_y\vec{j} + E_z\vec{k} = -\left(\frac{\partial V}{\partial x}\vec{i} + \frac{\partial V}{\partial y}\vec{j} + \frac{\partial V}{\partial z}\vec{k}\right) \tag{2.26}$$

where the part on the right of the equal sign is:

$$\nabla\vec{V} = \frac{\partial V}{\partial x}\vec{i} + \frac{\partial V}{\partial y}\vec{j} + \frac{\partial V}{\partial z}\vec{k} \tag{2.27}$$

Then:

$$\vec{E} = -\nabla\vec{V} \tag{2.28}$$

Formula (2.18) shows that $\nabla \cdot \vec{E} = \frac{\rho}{\varepsilon}$. It can be inferred from the formula that:

$$\nabla^2 V = -\frac{\rho}{\varepsilon_0} \tag{2.29}$$

If the electric field is in a dielectric, relative permittivity ε_r of the dielectric must be used in the formula as follow:

$$\nabla^2 V = -\frac{\rho}{\varepsilon_0\varepsilon_r} \tag{2.30}$$

This is the Poisson equation.

If there is free charge in the space under research, that is, $\rho = 0$, the preceding equation becomes:

$$\nabla^2 V = 0 \tag{2.31}$$

This is Laplace equation.

References

Crowe, C.T., J.D. Schwarzkopf, M. Sommerfeld, and Y. Tsuji. 2011. *Multiphase flows with droplets and particles.* Boca Raton: CRC Press.

Cundall, P.A., and O.D.L. Strack. 1979. A discrete numerical model for granular assemblies. *Gé otechnique* 29 (1): 47–65.

Di Renzo, A., and F.P. Di Maio. 2004. Comparison of contact-force models for the simulation of collisions in DEM-based granular flow codes. *Chemical Engineering Science* 59 (3): 525–541.

Kruggel-Emden, H., E. Simsek, S. Rickelt, S. Wirtz, and V. Scherer. 2007. Review and extension of normal force models for the Discrete element method. *Powder Technology* 171 (3): 157–173.

Maxey, M.R., and J.J. Riley. 1983. Equation of motion for a small rigid sphere in a nonuniform flow. The. *Physics of Fluids* 26 (4): 883–889.

Norouzi, H.R., R. Zarghami, R. Sotudeh-Gharebagh, and N. Mostoufi. 2016. *Coupled CFD-DEM modeling: Formulation, implementation and application to multiphase flows.* Newark: Wiley.

Ruspini, L.C., C.P. Marcel, and A. Clausse. 2014. Two-phase flow instabilities: A review. *International Journal of Heat and Mass Transfer* 71: 521–548.

Saffman, P.G. 1965. The lift on a small sphere in a slow shear flow. *Journal of Fluid Mechanics* 22 (2): 385–400.

Thornton, C., S.J. Cummins, and P.W. Cleary. 2011. An investigation of the comparative behaviour of alternative contact force models during elastic collisions. *Powder Technology* 210 (3): 189–197.

Part II
Basic Theory of Droplet Charging in Multiphase Flows

Chapter 3
Charging Ways and Basic Theories of Droplet Electrification

3.1 Charging Ways of Droplet Electrification

Droplet electrification is a phenomenon of industrial static electricity that exists in industrial liquid transport, injection, blending, stirring, filtering, mixing, spray coating and other processes. National Institute of Industrial Safety of Japan's Ministry of Health, Labor and Welfare, in the 1960s, did a research on 363 factories in various fields, such as chemical, pharmaceutical, and food processing. The research showed that there were 592 electrostatic charging processes, in which 234 processes involved droplet electrification, accounting for 39.5%. Among the 234 processes, there were 74 electrostatic charging processes during liquid transport, 27 injection processes, 37 blending processes, 23 stirring processes, 28 filtering processes, 14 mixing processes, and 31 liquid spray coating processes. Liquid transport and mixing processes would worth more attention among the preceding processes in industrial droplet electrification.

On the contact interface where the liquid phase is transforming into the gaseous phase or solid phase, ions form an electric double-layer. When this layer is split due to a mechanical effect, liquid is charged. The liquid dielectric, such as petroleum and other hydrocarbons, regardless of their purity, conductivity, contain a small amount of ionizable impurities. When the liquid is in the static state, impurity ions are selectively attracted to interior surfaces of the liquid container. But the opposite polarity ions with the equal quantity of electric charge will remain in the liquid phase and accumulate in the vicinity of the contact interface which form an electric double-layer. The ionic layer close to the solid surface is referred to as a fixed (charge) layer, and the opposite polarity ions layer in the liquid is called a diffusion (charge) layer. Figure 3.1 shows the electric double-layer.

© Springer Nature Singapore Pte Ltd. 2017
Z. Gu, W. Wei, *Electrification of Particulates in Industrial and Natural Multiphase flows*,
DOI 10.1007/978-981-10-3026-0_3

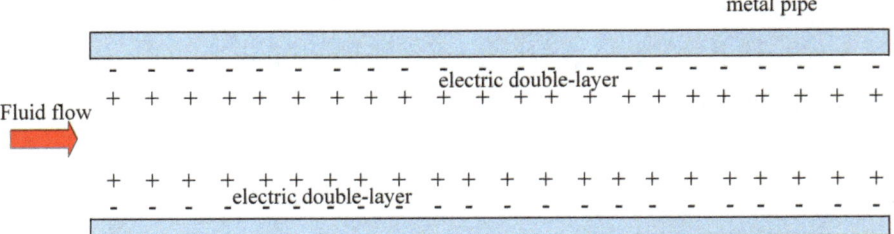

Fig. 3.1 Electric double-layer on the interface between pipe and fluid

3.1.1 Streaming Electrification

When liquid flow in a pipeline is driven by the pressure difference p between the two ends of the pipeline, it takes away the movable electric charge on the electric double-layer, making the liquid charged. This process is known as streaming electrification. When the liquid flow along the pipeline, ions on the diffusion (charge) layer are separated from those on the fixed (charge) layer and flow with the liquid, generating a charging current. The ions fixed on the interior pipeline surface are liberated due to the loss of ions in the corresponding diffusion (charge) layer. Part of these ions is neutralized during the recombination process, and the residual, excess charge is conducted to the earth along the pipeline surface.

1. Streaming current in conductive aqueous solutions

The value i_s of a streaming current varies with multiple factors, such as the liquid flow velocity, pipe diameter, and liquid properties. i_s is obtained using the global integral of space charge density ρ and flow velocity v over the electric double-layer, as shown in the following formula:

$$i_s = \pi d_p \int \rho v dx \qquad (3.1)$$

where

x is the distance from the pipeline interior surface (unit: m);
d_p is the pipeline diameter (unit: m).

Then, it can be deduced that the streaming current in a laminar flow can be calculated using the following formula:

$$i_s = -8\bar{v}\pi\varepsilon\xi \qquad (3.2)$$

where

$\bar{v}$ is the average flow velocity, m·s^{-1};
ε is the permittivity, F·m;

ξ is the voltage in V. It represents the potential difference between the fixed (charge) layer and diffusion (charge) layer on the electric double-layer.

$$\xi = \sigma \cdot \delta / \varepsilon$$

where

σ is the density of surface charge, $C \cdot m^{-2}$;
δ is the thickness of the electric double-layer, m;
ε is the permittivity, $F \cdot m^{-1}$.

If turbulence Re is between 2×10^3 and 10^5,

$$i_s = -0.040 Re^{3/4} \bar{v} \varepsilon \xi \tag{3.3}$$

where

Re is the Reynolds number that is obtained using formula $Re = \rho_m \bar{v} d_p / \eta$;
ρ_m is the fluid liquid density, $kg \cdot m^{-2}$;
η is the viscosity coefficient, $m^2 \cdot s^{-1}$.

Equations (3.2) and (3.3) show that, in a laminar flow regime, the streaming current is in proportion to the average flow velocity $\bar{v}$; when the turbulence Re is between 2×10^3 and 10^5, the streaming current is proportional to the $\bar{v}^{7/4}$. In the turbulence regime, the streaming current is affected more heavily by the flow velocity than in the laminar regime.

2. Streaming current in non-conductive solutions

Researches on the turbulence regime in non-conductive liquids are more practical and sensable. Thickness δ of the electric double-layer can be calculated using the following equation:

$$\delta = \sqrt{D_m \tau} \tag{3.4}$$

where

D_m is the molecular diffusion coefficient, $m^2 \cdot s^{-1}$;
τ is the relaxation time constant, s.

In non-conductive solutions, the thickness of the electric double-layer is greater than that in conductive solutions, affecting the streaming current in pipelines. In this regime, the streaming current can be calculated using the following formula:

$$i_s = \frac{\pi d_p \varepsilon \bar{v} \xi}{\delta} \tag{3.5}$$

Equation (3.5) shows that, in non-conductive solutions, the streaming current is affected by the thickness of the electric double-layer apart from the flow velocity $\bar{v}$. The value of the streaming current varies with the liquid conductivity.

3. Calculation method of the saturated streaming current

   ① Formulas used when the resistivity varies

If the liquid resistivity is greater than $10^{13}\Omega\cdot\text{cm}$,

$$I_\infty = 2.83 \times 10^{-15} \varepsilon_r v^{1.875} d_p^{0.875} \eta^{-0.625} \tag{3.6}$$

If the liquid resistivity is less than $10^{13}\Omega\cdot\text{cm}$,

$$I_\infty = 1.1 \times 10^{-25} \varepsilon_r^{1.5} v^{2.83} d_p^{0.625} \eta^{-1.375} \rho^{0.5} \tag{3.7}$$

   ② Generic formula:

$$I_\infty = Av^\alpha d_p^\beta \tag{3.8}$$

where

A is the calculating coefficient. It equals to 3.75×10^{-6}, $\text{s}^2\cdot\text{m}^{-4}$ if the hydrocarbon liquid, such as kerosene and gasoline, flows in long and straight pipelines.
α and β are coefficients determined by the pipeline diameter.

4. Factors related to streaming electrification in liquids

   ① Impurities

The liquid electric polarity also varies with impurity properties. The liquid contains impurities with high molecular materials (such as rubber and asphalt) would increase the generation of static electricity. When a liquid contains moisture, additional static electricity will also be generated during liquid flowing, stirring, and injection processes. Colloidal particles in liquids attract free ions and thus being charged. If the permittivity of colloidal particles is greater than the liquid permittivity, the colloidal particles are positively charged, and otherwise negatively charged. Droplets in oil products can be regarded as colloidal particles that are positively charged during settlement (The range that oil products are mixed with 1–5% moisture is extremely dangerous).

   ② Liquid conductivity (resistivity)

The liquid charge increases with increasing resistivity within a certain resistivity range but decreases with increasing resistivity beyond the resistivity range. Tests have shown that the liquid with a resistivity of $10^{11}\Omega\cdot\text{m}$ is easily charged but not easy to be charged with a resistivity below $10^8\Omega\cdot\text{m}$ and above $10^{13}\Omega\cdot\text{m}$.

③ Pipeline materials and roughness levels of interior surfaces

Pipeline materials affect the liquid charge through electrostatic dissipation rather than material electrification performance. A rougher pipe interior surface means a greater contact area and more streaming and separation, resulting in a greater streaming current and higher charge degree.

④ Liquid flow condition

When a liquid changes from a laminar flow to turbulence, the liquid carries static electricity. Because of the change in the fluid flow state, new charges are generated due to the liquid thermal motion and collision. The velocity distribution changes when the liquid flow regime changes from laminar flow to turbulent. When the liquid is a laminar flow, the liquid velocity is parabolically distributed along the pipeline diameter. When the liquid is a turbulent, the liquid velocity is uniform in the middle of the pipeline, despite that the velocity gradient is greater than that of the laminar flow. The change in the velocity gradient causes more charges on the diffusion layer to converge to the pipe center, resulting in a greater liquid charge and an increase in the charge density of the entire pipe when compared to the laminar flow.

3.1.2 Settlement Electrification

When particles are suspended in the liquid settle, the particles and liquids carry charges with different properties, producing a potential difference in the upper and lower parts of the liquid container. The potential difference is known as the settlement electrification.

Settlement electrification can also be explained with the electric double-layer theory. When solid particles are present in the liquid, an electrical double-layer is formed on the solid-liquid contact interface. If the solid particles settle, they take away the charges adsorbed on their surfaces, making the liquid and solid particles charged with different polarities. An electrostatic field is then generated in the liquid, and a potential difference occurs in the upper and lower parts of the liquid.

3.1.3 Injection Electrification

Liquid injection electrification occurs when liquid are ejected from a nozzle at a high speed. In this case, the nozzle and droplets carry charges with opposite polarities. The electric double-layer theory can also explain the cause of this kind of electrification. Rapid contact and separation occurs between the nozzle and droplets when liquid are injected from a nozzle at a high speed. When the nozzle and liquid droplets contact, the droplets take away charges with one polarity, leaving charges with the other polarity. As a result, liquid droplets and the nozzle carry charges with opposite polarities.

In addition, when the liquid is ejected from the nozzle under a high pressure, it forms a jet flow and is then split into small droplets when contacting with the air. Relatively large droplets quickly settle while small droplets remain in the air, forming a mist-like droplet cloud. This cloud is a charge cloud with a large amount of charges.

3.1.4   Liquid Impingement Electrification

When a liquid is ejected from a pipeline to a wall or a board, the liquid splashes upwards into tiny droplets that become charged when split. These charged split droplets form a charge cloud. This electrification way is common in the storage and transport of oil products, for example as light oil. When light oil is injected from the top inlet of an oil tank or a tank car, the oil column impacts the tank interior surface or oil surface and produces splashing droplets, bubbles and mist droplets, generating static electricity.

3.1.5   Splash Electrification

When a liquid is splashed or poured onto a non-infiltrated solid, splashing droplets roll on the solid surface, causing the solid to carry charges with one polarity and the liquid to carry charges with the other polarity. This phenomenon is known as splash electrification. When the droplets fall onto the solid surface, an electric double-layer is formed on the contact interface. The droplet inertia drives the droplets to roll on the solid surface continuously. Droplets take away charges on the diffusion layer and become charged with one polarity while charges on the fixed layer stay on the solid surface with the other polarity.

3.1.6   Interface Electrification Between Gas-Liquid

Generally, there are two types of droplet electrification: natural electrification caused by liquid flow, split, evaporation and so on; and forced electrification caused by charging by a power supply, ion source and other outside energy.

Electrification caused by droplet split has attracted researchers' attention since 1890 when Elster, Geiter, and other people observed intense electrification phenomenon of a waterfall that generated electrostatic sparks. Then, Lenard investigated waterfall electrification and found that small particle-size droplets carried negative charges while large particle-size droplets carried positive charges. This symptom was believed to occur because, on the contact interface between gas and liquid, the negative charges on the electric double-layer within the liquid surface are split.

On the contact interface between droplets and the air, polar molecules are aligned. As the result, these polar molecules attract a lot of negative ions close to the liquid surface but in the liquid and a slight amount of positive ions. These polar molecules can move freely in the liquid, forming an electric double-layer in the liquid. When the small droplets are split, fission droplets with a smaller particle size would carry negative net charges.

There are many ways for liquid forced electrification, such as:

- Immerse a metal wire electrode in the liquid and apply DC voltage to the electrode, to electrify droplets at the container bottom;
- Use corona discharge or field emission to electrify droplets;
- Use ultraviolet, X-ray and other high-energy electromagnetic waves to radiate metal electrodes in the liquid so that the metal electrodes release electrons. Then, inject liquid to make droplets charged;
- Use induction charges or radioactive material charges.

3.1.7 Diffusion Electrification

The processes responsible for the electrification of thunder clouds are apparently centered within the mixed phase zone where water droplets, ice crystals, graupels may coexist. These hydrometeors with different weight and structure may acquire different charges under polyphase interfacial interactions which lead to mass transfer (along with charge transfer) or just charge transfer. Many observational and simulation studies of storms have also supported the hypothesis that graupel and ice crystals are the species that play the most vital role in the electrification of storms and occurrence of lightning.

Thundercloud electrification is typically characteristic of tripole charge structure, which has been elucidated by current models with the recognized relative growth rate theory involved. This theory was firstly proposed by Baker et al., suggested that: "During ice particle collisions, the particle whose surface is growing fastest from vapour diffusion at the instant of collision will charge positively on particle separation".

In the lower (warmer) region, the surface of a riming graupel would grow directly from the vapour in the cloud as well as from the vapour supplied by droplets freezing upon its surface. The time required for freezing is longer at warmer temperatures. During this time the surrounding ice surface would grow rapidly from the vapour supplied by the freezing water. Hence, the surface of a riming graupel would grow faster from the vapour at warmer temperatures and also at higher liquid water contents, and it would be charged positively according the fast growth theory. In the higher (colder) region, the ice crystal may be growing faster on the average and charges positively, while the graupel would charges negatively at colder temperature. In cloud, the graupel pellet will be warmer than the ice crystals by release of latent heat from accreted droplets; given a high enough accretion rate, the rimer surface temperature approaches 0 °C. Experiments conducted by Saunders et al. showed that the effect of heating a riming target sufficiently to reduce its rate of growth by vapour deposition below that of the rebounding ice crystals resulted in the graupel target acquiring negative charge. So, during crystal/graupel interactions, the ice crystal is growing faster at colder temperature is reasonable and the riming graupel charges negatively even if it is the warmer of the interacting particles.

3.2 Basic Theories of Droplet Electrification

3.2.1 Contact Charging

1. Equivalent electrical circuit model of contact charging

Owing to the direct contact of the metal electrode with the liquid flow in contact charging, the polarization occurs in the nozzle. The liquid flows through the nozzle would be atomized under the joint action of mechanical force and Coulomb force. The columnar liquid between the nozzle and the electrode is equivalent to a resistance, R, when connected to the DC voltage. The charging system can be regarded as an active circuit. The DC voltage of electrostatic power supply has a voltage, E, and an internal resistance R_1. Once the electric ring is grounded, the nozzle has a potential voltage, U_0, and the ground potential of the ring electrode is zero, thus the equivalent circuit is shown in Fig. 3.2.

According to the equivalent circuit of contact charging, the current flow through the liquid can be expressed as:

$$i = \frac{E}{R + R_1} \tag{3.9}$$

2. Space potential induced by contact charging

In the contact charging configuration, the distribution of the space potential conforms to the Laplace equation if the axis of the nozzle coincident with that of the ring electrode. Setting the nozzle radius, r_1, the radius of ring electrode, r_2, the definite solution under polar coordinates is as follows:

Fig. 3.2 The equivalent
circuit of contact charging

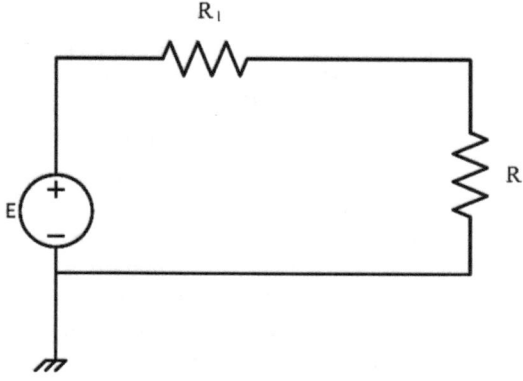

$$\begin{cases} \nabla^2 U = 0 \\ U\big|_{z=0} = \begin{cases} U_0 & 0 \le r \le r_1 \\ U_1(r) & r_1 \le r \le r_2 \\ 0 & r_1 \le r \le r_2 \end{cases} \end{cases} \qquad (3.10)$$

Where, $U_1(r)$ is the potential in xoy plane, it still meets the Laplace equation. The solution is:

$$U_1 = \frac{U_0}{\ln \frac{r_1}{r_2}} \ln \frac{r}{r_2} \qquad (3.11)$$

Solving Eq. (3.10) as follows:

$$U = \frac{1}{2\pi} \int_0^{2\pi} \int_0^{r_1} \frac{z r U_0}{[x^2 + y^2 + z^2 + r^2 - 2r(x\cos\theta + y\sin\theta)]^{\frac{3}{2}}} \, d\theta dr$$

$$+ \frac{1}{2\pi} \int_0^{2\pi} \int_{r_1}^{r_2} \frac{z r U_0 \ln \frac{r}{r_2}}{[x^2 + y^2 + z^2 + r^2 - 2r(x\cos\theta + y\sin\theta)]^{\frac{3}{2}} \ln \frac{r_1}{r_2}} \, d\theta dr \qquad (3.12)$$

3. Dielectric polarization of the flow jet

In the electrospray system, both the droplet charging process and the selecting of electrical field strength are concerned with the dielectric polarization. The dielectric polarization behavior is usually characterized by the relative dielectric constant, ε_r. The dielectric capacity is related to its own molecular structure. For the dielectric composed of polar molecules, polar molecules in the dielectric would have an alignment under the action of an external electric field; while for the dielectric composed of nonpolar molecules, the positive and negative charges of molecules shift each other along the opposite direction of the electric forces under the action of the applied electric field, resulting in the polarization of positive and negative charges, eventually the macroscopic polarization effect of the dielectric. For the composite of different dielectrics, under the action of the applied electric field, free charge carriers would accumulate in the defects of material and the interface between different dielectrics, and forms the local accumulation of space charge, which is known as the space charge polarization or the interface polarization between different dielectrics.

For any dielectric, the constituent dipole moment caused by external electric fields is related to not only the externally applied electric field intensity, but also the internally induced field strength of dipole moment within the dielectric. That is to say, the dielectric field strength at any point within the dielectric, the effective field E_e, is the coupling effect of the externally applied electric field and the induced electric field formed by the polarization charge. The induced dipole moment of dielectric per unit volume, or the polarization vector, $\vec{\beta}$, is:

$$\vec{\beta} = \lim_{\Delta V \to 0} \frac{\sum_{i=1}^{n} \vec{P}_i}{\Delta V} \tag{3.13}$$

Where $\vec{P}_i$ is the constituent electric dipole moment, along the direction from negative to positive charge; n is the number of constituent particles; ΔV is infinitesimal volume. $\vec{P}_i$ is the product of the charges q and the charge centroid distance, L, between the positive and negative charges, namely:

$$P = qL \tag{3.14}$$

Obviously, $\vec{P}_i$ has the direct relations with the effective electric field intensity, $\vec{E}_e$, usually expressed as

$$\vec{P}_i = \alpha \vec{E}_e \tag{3.15}$$

Where, α represents the microscopic polarizability. If the number of constituent particles per unit volume, N is known,

$$\vec{\beta} = N\alpha \vec{E}_e = \chi \vec{E}_e \tag{3.16}$$

Where, χ represents the polarization coefficient.

The dielectric polarization under the action of the applied electric field generates the induced electric dipole moment, and the appearance of the polarization charges on the surface of the dielectric. The strength of polarization would also be characterized by the polarization charge capacity on the surface of the dielectric. For the charging in the applied electrostatic field, the pressurized liquid jet flows out of the nozzle and then breaks up into the charged droplets. The atomization mechanisms are related to the flow rate, pressure and physical properties of liquid jet. Under different operation conditions, the breakup types of liquid jet have dropwise breakup, filament-wise breakup and film-wise breakup (Jaworek and Sobczyk 2008; Raizer et al. 2011). However, irrespective of any breakup type, the flow jet has come into polarization before its breakup, which is a kind of the polarization of polar liquid. In the applied electric field, the polarizations involve the polarization of electron shift, the polarization of dipole steer, and the interface polarization between the flow jet and the surrounding gas. Due to the inhomogeneity and complexity of the electrospray field, irregularity and diversity of flow jet surface at the outlet of the nozzle, it is very difficult to accurately calculate the polarization intensity. Approximately, assuming that the normal direction of flow jet surface at the outlet of the nozzle is parallel to the direction of the electric field, the polarization intensity of flow jet can be expressed as

$$\beta = \sigma_j \tag{3.17}$$

Where, σ_j represents the polarization charge density on the flow jet surface at the outlet of the nozzle.

If there is an angle between the direction of the applied electric field and the normal direction of dielectric jet surface, θ, therefore, $\beta = \sigma' \cos \theta$.

And because

$$\sigma' = \varepsilon_0(\varepsilon_r - 1)E = (\varepsilon - \varepsilon_0)E \tag{3.18}$$

Then, Clausius equation is obtained as

$$\beta = (\varepsilon - \varepsilon_0)E = N\alpha E_e \Rightarrow \varepsilon = \varepsilon_0 + N\alpha \frac{E_e}{E} \tag{3.19}$$

Set $\varepsilon_r - 1 = x$, so called the polarization coefficient (or electric susceptibility), there is

$$\beta = x\varepsilon_0 E_e \tag{3.20}$$

It indicates that the polarization intensity presents a linear relationship with the applied electric field. In a determined electric field intensity, the polarization intensity increases with the relative dielectric constant, ε_r.

4. Conductivity of the jet flow

The electric conductivity of the polarity jet flow is larger than the nonpolar jet flow, especially, the impurities in the jet flow have great influence on the conductivity of the jet flow. There is only a little part of polar molecules and impure molecules in the jet flow dissociating into ions, which are involved in conduction. At an ambient temperature, when the applied electric field strength much less than the breakdown strength of the surrounding gas, the conductivity of impurity ion γ has no relationship with the applied electric field strength, the conducting current in the jet flow follows the Ohm's law. In this situation, the electric conductivity are determined by the ionic charge, q, the average substeps, δ, the concentration of ions n_0, the dissociation barrier u_a, and the average barrier to overcome in the transition of electron or ion, u_0, the number of dissociated molecular per second in unit volume, N, the temperature T (absolute temperature), the vibration frequency of ions, ν, the relative thermal vibration frequency, ν_0. The relationship can be expressed as

$$\gamma = \frac{q\delta\nu}{6kT}\sqrt{\frac{N\nu_0}{\xi}}e^{-\frac{(2u_a+u_0)}{2kT}} \tag{3.21}$$

where, ξ is the recombination coefficient of ions; k is the Boltzmann constant, equal to the ratio of the universal gas constant and Avogadro constant.

It could be concluded from the above theoretical analysis that for electrospray, the resistivity of the jet flow is related to the physical properties of liquid jet and the temperature of jet flow. Under the same conditions of operation environment and liquid jet, once the contact charging voltage is less than the breakdown voltage of the surrounding gas, the volt-ampere characteristic of jet flow can be described by the Ohm's law, meaning the charges on droplets are proportional to the voltage between electrodes.

3.2.2 Induction Charging

1. Equivalent electrical circuit model of induction charging

The pressurized liquid jet flow, passing through the nozzle at an axial velocity, v (m/s), forms the cone jet near the outlet of the nozzle. And then the cone jet, impacting with the surrounding gas, breaks up into a cloud of droplets. Induction charging is implemented at the front of the cone jet where droplets form, resulting into further breakup of the droplets in motion. The electrode exerted on the cone jet is usually a ring metal electrode, the opposite electrode is a plate electrode where the droplets finally deposit, namely the liquid electrode. The surrounding gas (air) between the metal electrode and the liquid electrode isolates the droplets from the metal electrode. Once a high voltage electric field is exerted on the ring electrode, the potential difference between the applied electric field of the ring electrode and the cone jet charged the droplets with induction charging. The availability of induction charging is determined by the distance between the cone jet and the metal electrode. The implementation of this induction charging must maintain the insulation of jet flow from the metal electrodes. The surrounding air layer becomes the insulating medium between the metal electrode and the liquid electrode. The metal electrodes, the liquid electrode and the surrounding air layer are composed of an air capacitor, which presents a large resistance in DC. Thus, the circuit for induction charging would be a resistance capacitance parallel connection circuit. The equivalent circuit of induction charging process is illustrated in Fig. 3.3.

Fig. 3.3 The equivalent
circuit of induction
charging

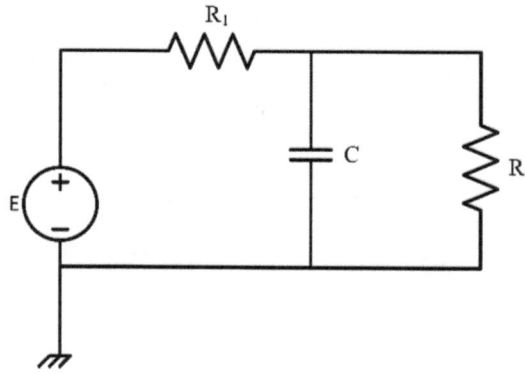

2. Space potential induced by induction charging

In order to get the same polarity of the charged droplets like contact charging, the ring in induction charging must be connected to the opposite electrode to that in contact charging. Setting the potential of ring as high negative voltage, U_0, and the nozzle grounded with zero potential, the space potential of the ring electrode also conforms to the Laplace equation, its definite solution under polar coordinates is as follows:

$$\begin{cases} \nabla^2 U = 0 \\ U\big|_{z=0} = \begin{cases} U(r) & r > r_1 \\ 0 & r \le r_1 \end{cases} \end{cases} \tag{3.22}$$

The solution satisfies the Green's function on the half space conditions, thus the space potential is:

$$U = \frac{1}{2\pi} \int_0^{2\pi} \int_{r_1}^{r_2} \frac{zrU_0 \ln \frac{r}{r_1}}{[x^2 + y^2 + z^2 + r^2 - 2r(x\cos\theta + y\sin\theta)]^{\frac{3}{2}} \ln \frac{r_2}{r_1}} d\theta dr \tag{3.23}$$

The space potential is the sum of the potential of the ring electrode and the potential caused by charged droplets. The charged droplets still keep charged even far away from the nozzle. The potential effect of the ring electrode lessens with the increase of the distance from the ring electrode. At the depositing zone of droplets, the electric field generated by the ring potential is low enough to take little effect on the induction charging of droplets. Thus, the potential at the depositing zone can be regarded as the production of the charged droplets.

Along the same axis of the nozzle, the cone jet breaks up, forming the symmetric distribution of the charged droplets. According to Eq. (3.12), the potential of contact charging on the axis is as follows

$$U = \frac{zU_0}{\ln \frac{r_1}{r_2}} \left[\begin{array}{l} \dfrac{1}{2z^2 r_1} + \dfrac{1}{4z^3} \ln \dfrac{\sqrt{z^2 + r_1^2} - z}{\sqrt{z^2 + r_1^2} + z} - \dfrac{1}{2z^2 r_2} - \\[2ex] \dfrac{1}{4z^3} \ln \dfrac{\sqrt{z^2 + r_2^2} - z}{\sqrt{z^2 + r_2^2} + z} - \dfrac{\ln \dfrac{r_1}{r_2}}{2\sqrt{(z^2 + r_1^2)^3}} \end{array} \right]$$

$$+ U_0 \left(1 - \frac{z}{\sqrt{z^2 + r_1^2}} \right) \tag{3.24}$$

From Eq. (3.23), the potential of induction charging on the axis of the cone jet can be expressed by:

$$U = \frac{zU_0}{\ln \frac{r_2}{r_1}} \left[\begin{array}{l} \dfrac{1}{2z^2 r_1} + \dfrac{1}{4z^3 r_1} \ln \dfrac{\sqrt{z^2 + r_1^2} - z}{\sqrt{z^2 + r_1^2} + z} - \dfrac{1}{2z^2 r_2} - \\[2ex] \dfrac{1}{4z^3} \ln \dfrac{\sqrt{z^2 + r_2^2} - z}{\sqrt{z^2 + r_2^2} + z} - \dfrac{\ln \dfrac{r_2}{r_1}}{2\sqrt{(z^2 + r_2^2)^3}} \end{array} \right] \tag{3.25}$$

According to Eqs. (3.24) and (3.25), the potential of contact charging declines sharply with the increase of the distance between the jet spray and the nozzle, while the potential of induction charging declines more slowly.

3. Conductivity of the gas dielectric in induction charging

The surrounding gas is used as the insulating medium in electrospray, so the conductive properties of the gas medium need to be analyzed firstly. Whether the gas medium composed of non-polar molecules or polar molecules, the gas has good electric insulation owing to its large molecular space, small density and molecular interactions under ambient temperature and atmospheric pressure. The effective electric field intensity in the gas medium is equal to the average of the macroscopic electric field intensity, namely, $E_e = E$. However, a small amount of ions in the gas, produced by thermal motion, light and radiation, have the directional migration under the action of the applied electric field, resulting in the presence of the weak electrical conductivity of the gas. The electrical conductivity of the gas under atmospheric pressure is shown in Fig. 3.4. The electrical conductivity of the gas can be divided into three stages. In stage I, the electric field strength is small, less than the E_1, the electric current is proportional to the electric field intensity. The conductive current of the gas follows the Ohm's law. In stage II, the electric field intensity is larger than E_1 but less than E_2, the conductive current in the gas is saturated and keep unchangeable even if increasing the electric field intensity, the saturation current depends on the production of the carrier in the gas. The stage III comes, characteristic of the current surging when the electric field strength is larger than E_2. In stage III, the conductive particles in the gas get enough energy under the action of the high electric field intensity, and eventually become ionized when colliding with other molecules. A sharp increase of the carriers in the gas causes the

Fig. 3.4 The gas conductive volt-ampere characteristics under the applied voltage

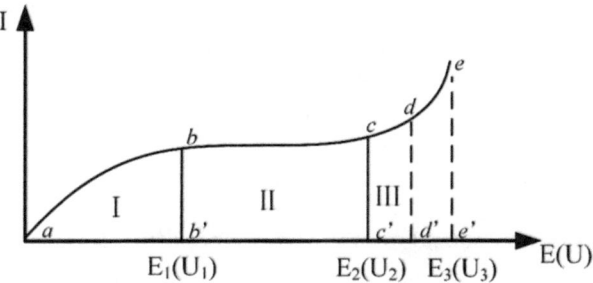

exponentially growing of the conductive current, and consequently the breakdown of the gas. After the breakdown of the gas, the electric field between the electrodes decreases.

According to the gas conductive volt-ampere curve, the conductivity of gas would follow the Ohm's law under the low exerted electric field. The volume conductivity, g, can be calculated by:

$$g = Nq\eta \tag{3.26}$$

Equation (3.26) show the relationship between the macroscopic parameter, g, and the gas dielectric microcosmic parameters of the unit volume carrier number N, the carrier charge, q and the carrier mobility, η. This equation can also be applied to calculate the conductivity of liquid, solid dielectric.

For the liquid dielectric between the electrodes, the change of the current within the liquid dielectric with the applied electric field is similar to that of the gas dielectric. Therefore, the current within the liquid dielectric obeys the Ohm's law in the scope of the low applied electric field, and the carrier within the liquid dielectric contributes to the current. However, as the viscosity of the liquid dielectric is higher than that of the gas dielectric, the relationship between the carrier mobility and the liquid viscosity is as follows:

$$\mu\eta = \frac{e}{6\pi r} \tag{3.27}$$

Where r is the stokes radius of the carrier, and e is the charge of the carrier.

With the increase of electric field intensity, for the liquid dielectric, especially for impure liquid medium, the three stages of the electric conductivity are difficult to identified, roughly similar to that of the gas electric. The carriers within liquid dielectric are nearly composed of the small amount of polar molecules and the dissociated ions of the impure molecules. Based on the theory of reaction rate, the ion density n related to the absolute temperature T is shown as follows:

$$n = n_0 e^{-E_n/bT} \tag{3.28}$$

Where, n_0 stands for a constant independent of the temperature, b is the Boltzmann constant, and E_n stands for the work of an electrolyte molecule dissociating into positive and negative ions.

The conductivity of liquid dielectric is:

$$g = g_0 e^{-E_\beta/bT} \tag{3.29}$$

Where, E_β stands for the conductance activation energy, which equals to the sum of the activation energy of viscous liquid, E_μ and the activation energy for ion mobility, E_η.

Generally, the electrical conductivity of the liquid dielectric and the gas dielectric is related to their physical properties, and the temperature in electrospray. The conductive property obeys the Ohm's law under the low electric field intensity, and the charge density on droplets is proportional to the applied voltage.

4. Capacitance of the electrode

The electrode for the polarization of the droplets from the breakup of the flow jet out of the nozzle would be a single plate electrode, or double plate electrodes if intending to improve the charging efficiency. The double plate electrodes are implemented to obtain the atomization of the jet spray and the charging of the droplets between the two plates. For single plate electrode, the capacitor is composed of a single electrode plate and the fan-shaped region distributed by the cloud droplets. For double electrode plates, the capacitor is a kind of shunt capacitor composed of double plate electrodes and the fan-shaped region distributed by the cloud droplets. Setting the area of the plate electrode as A, the distance between the plate and the fan-shaped region d, and ignoring the edge effect, the capacitance is:

$$C = \frac{\varepsilon A}{d} \tag{3.30}$$

Setting the voltage as U, the induction charge of the fan-shaped region distributed by the cloud droplets is:

$$q = CU = \frac{\varepsilon A}{d} U \tag{3.31}$$

Obviously, the charge density of the induction charging in the fan-shaped region distributed by the cloud droplets grows with the increases of the applied voltage and the plate area, but with the decrease of the distance between the plate electrode and the fan-shaped region.

The electrode for the conical spray from the nozzle is often ring electrode. The capacitor composed of the ring electrode and the cone liquid spray is simplified as the gas dielectric capacitor composed of the coaxial cylindrical electrodes. Setting the radii of the inner cylinder and the outer cylinder are r_1 and r_2, respectively, and setting the applied voltage as U, the capacitance of the coaxial cylindrical capacitor is as follows:

$$C = \frac{2\pi\varepsilon}{\ln \frac{r_2}{r}} \tag{3.32}$$

The charge density of the capacitor can be expressed as:

$$q = CU = \frac{2\pi\varepsilon U}{\ln \frac{r_2}{r_1}} \tag{3.33}$$

The charge density is closely related to the dielectric constant of the dielectric between the plates, ε. The larger the dielectric constant ε, the greater the charge density is. The capacitor of the flow jet dielectric between the induction charging electrode and the nozzle is regarded as the coaxial capacitor. The charge density is large in practice, since the dielectric constant of the flow jet dielectric is much more than that of the surrounding gas.

For the coaxial capacitor within the same dielectric, exerted by the same electric voltage, the charge density on the plate is only related with the characteristic scale of the capacitor, the inside radius, r_1, and the outside radius, r_2 of the coaxial cylindrical capacitor. The smaller the scale ratio of r_2 to r_1, the more charge density of the coaxial capacitor is. In practice, the adjustment of the charge density of the coaxial capacitor would be reached by changing the radius and the clearance between the two electrodes, which can be expressed as:

$$
\begin{aligned}
dq &= \frac{\partial q}{\partial r_2} dr_2 + \frac{\partial q}{\partial r_1} dr_1 \\
&= \frac{2\pi\varepsilon U}{(\ln r_2 - \ln r_1)^2} \left(\frac{1}{r_1} dr_1 - \frac{1}{r_2} dr_2 \right)
\end{aligned} \tag{3.34}
$$

Keeping the outside radius, r_2 constant, the radius of the cone spray liquid electrode, r_1, would be adjusted by shifting back and forth the installation position of the ring electrode relative to the nozzle. Theoretically, if the radius, r_1 is close to r_2, the amount of charge density reaches the maximum of charge density, the induction charging gets the best efficiency. The clearance between the two electrodes, δ, which is equal to the radius difference, $r_2{-}r_1$, is changeable due to the instability of the jet spray. If the clearance between the two electrodes is too small, it could easily cause the cone jet contacting with the electrode in operation. The induction charging becomes into the contact and the polarity of charge on droplets would be changed.

When $\delta/r_2 \ll 1$, $\ln\frac{r_2}{r_1} \approx \frac{\delta}{r_2}$, the charge density can be expressed as:

$$q = CU \approx \frac{2\pi \varepsilon r_2 U}{\delta} \tag{3.35}$$

$$dq = -\frac{2\pi \varepsilon r_2 U}{\delta^2} d\delta \tag{3.36}$$

Equation (3.29) shows that the charge density is proportional to the dielectric constant of the cone liquid spray and the applied voltage on the electrode in induction charging, and inversely proportional to the clearance between the electrodes.

3.2.3 Corona Charging

1. Equivalent electrical circuit model of corona charging

As shown in Fig. 3.4, in the current surge stage of the gas breakdown conduction, when the applied voltage up to the point of d', not only the negative ions with high mobility collide with the neutral molecules and make the neutral molecules ionized with the neutral molecule, but also the positive ions with low mobility, colliding with the neutral molecules, make the neutral molecules ionized. Consequently, new ions are generated under the applied electric field. Corona, the aureole in the ionizing zone around the electrode, is the corona discharging from the ultraviolet radiation produced by the excited molecules. The energy of the excited molecules comes from the collision of electrons and neutral molecules. The ionization current through the gas is the corona current. The starting voltage at the point of d' is the critical corona voltage, thereafter, the corona charging dominates. While the applied voltage increases up to e', all the dielectric between the electrodes comes into breakdown, resulting in a short discharging current and then sharply decreasing of the applied voltage. With the same applied voltage, the current produced by the negative corona is larger than that of the positive corona, the divergence of the positive corona discharge is different from the convergence of the negative corona discharge, and the field strength of the negative corona breakdown is higher than that of the positive corona. For the spray liquid of polymer compound composed of neutral molecules, molecules through gaining electron becomes negative more easily than molecules through losing electron becomes positive, therefore, it is conducive to use the negative corona charging for cone jet spray.

The corona charging is the ionization of the air around the cathode tip under the applied voltage makes the droplets charged, the resistance, Ra, between the cathode and the cone jet reduces quickly with the strengthening of the ionization. A variable resistor, instead, is used in the equivalent circuit diagram of the corona charging, as shown in Fig. 3.5.

Fig. 3.5 The equivalent circuit of corona charging

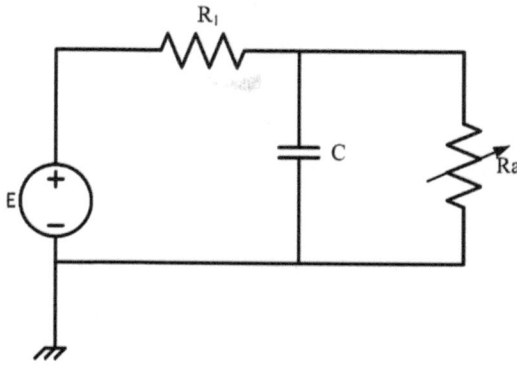

2. Space potential induced by corona charging

Since the size of corona cathode much less than the distance between the cathode and the earth electrode and the corona discharging takes place in the small region around the cathode tip, the corona cathode would be regarded as a point charge. Neglecting the tangential component of the electric field intensity around the cathode, the cathode is also regarded as an equipotential body of conductor, with the potential, U_0 and the charge density, q_0'.

At the earth boundary of the electric field induced by the point charge of the corona charging around the cathode, the point charge of the corona cathode would also induce the charge on the ground surface. Generally, with no more spray of the cone jet in the corona charging, the electric field between the corona electrode and the grounding electrode is the sums of the point charge induced electric field and the applied electric field. Irrespective of the distribution of the induced charge on the ground surface, the potential on the ground surface is 0, therefore the Laplace equation could be used to analyze the space potential, U.

$$\begin{cases} \dfrac{\partial^2 U}{\partial x^2} + \dfrac{\partial^2 U}{\partial y^2} + \dfrac{\partial^2 U}{\partial z^2} = 0 \quad \text{(except the point charge)} \\ U|_S = 0 \end{cases} \tag{3.37}$$

The induced charge in corona charging is simply equivalent to the point charge. Based on the principle of uniqueness, the equivalent charge is the mirror image of the origin charge if the ground surface likes a mirror, keeps the charge distribution and the potential, 0.

In the area of $x \geq 0$ and $z \geq$ '0, the space potential induced by the point charge can be solved as follows.

$$U(x, y, z) = \frac{q_0'}{4\pi\varepsilon_0}\left(\frac{1}{r_1} - \frac{1}{r_2}\right)$$

$$= \frac{q_0'}{4\pi\varepsilon_0}\left[\frac{1}{\sqrt{x^2 + y^2 + z^2}} - \frac{1}{\sqrt{x^2 + (y+h)^2 + z^2}}\right] \tag{3.38}$$

3. Conductivity of the ionized gas dielectric

The degree of the gas ionization in corona discharge is confined. Small part of the gas molecules would be ionized into electrons and ions, meaning the weak ionization the gas, besides, the other gas molecules still keep the neutral state. Actually, only 1% for the ionization degree could almost realize the conductivity of all the ionized gas molecules. Spitzer (1962) gave the conductivity of fully ionized plasma:

$$\gamma_c = \frac{1.56 \times 10^{-4}T^{\frac{3}{2}}}{\ln\left(1.23 \times 10^{-4}T^{\frac{3}{2}}n_e^{-\frac{1}{2}}\right)} \quad \text{(V/cm)} \tag{3.39}$$

Where, T is the absolute temperature; n_e is the electron density in the ionized gas dielectric.

The collisions between the particles at any degree of the gas ionization are different from those between the fully ionized gas dielectric. The plasma at any ionization degree is composed of cation, anion, electron and neutral molecule (atom). Under the action of the applied electric field or the gradient of the pressure and temperature field, the ions in the plasma will move in the same direction to the electric field or in the direction of pressure and temperature decreasing, leading to the diffusion current of charge motion. The theory of Chapman-Cowling theory determines the conductivity of the weak ionization (Chapman and Cowling 1958)

$$\gamma_w = 3.34 \times 10^{-12}\frac{\alpha}{ST^{\frac{1}{2}}} \tag{3.40}$$

Where, S is the cross section area of collision, T is the absolute temperature and a is the ionization degree.

In fact, the full ionization and the weak ionization are the two extreme circumstances of the gas ionization, and in general, for any degree of the gas ionization, the conductivity parallel model is often employed to calculate the conductivity.

4. The current and voltage in corona charging

The initial existence of electrons is necessary to produce the corona discharge, since the gas dielectric is composed of all neutral particles, with no electrons, the ionization or discharge could not take place regardless of high voltage between the electrodes. Under the action of the exerted electric field, the free electrons colliding with neutral molecules would cause a chain of the ionization of the neutral molecules and consequently the avalanche growing of electrons in the ionized gas dielectric. The ionized gas finally comes into the conductive medium. The current, I, exponentially increases with the distance from the corona charging electrode, x (Jung et al. 2011)

$$I = I_0 e^{\alpha x} \tag{3.41}$$

where α is the Townsend discharge coefficient, dependent on the pressure, P and the electrical field intensity, E, satisfying $\frac{\alpha}{P} = A e^{-\frac{BP}{E}}$, where A and B are both constants.

When the applied voltage between the electrodes rises up to the critical value, the gas discharging takes place between the electrodes. The instantaneous voltage at the start of the discharging is called as the firing voltage or discharge inception voltage. Thompson analyzed the electronic multiplier effect of ionization (α function), the ionization produced by the ion collisions with gas molecules (β function) and the bombardment effect of the accelerated ion impacting on the cathode under the applied electric field (γ function) and proposed the Townsend spark discharge conditions (Wadhwa 2007)

$$\gamma \left(e^{\int_0^l \alpha dx} - 1 \right) = 1 \tag{3.42}$$

Where, α and γ represent the ionization coefficient of α function and γ function. l is the electrode spacing.

Paschen stated that the ignition voltage is dependent on the product of the pressure, P and the electrode spacing, l, determined by the minimum product and expressed as (Wadhwa 2007)

$$U_s = \frac{BPl}{\ln\left(APl/\Phi\right)} \tag{3.43}$$

Where, A and B is the constants, respectively; $\Phi = \alpha l$ is dependent on the Townsend's condition.

The electric field intensity, E_x at any point, x in the applied electric field is a function of the applied voltage U, which determines the discharging start voltage and the electric field intensity for different geometric shape of the electrode. In addition, the surface property of the electrode has a lot to do with the corona voltage as well.

After the gas dielectric coming into the spark discharge, the potential between the two electrodes would have a "sudden change", as a result of the avalanche growing of electron in the gas dielectric, as shown in Fig. 3.6. At the start of spark discharge, t = 0, the gas would just be broken down, the voltage between the two electrodes is the initial voltage, U_0, and the potential in the gas dielectric rises in the dotted straight line. As the spark discharge goes, a lot of ions and electrons owing to the electronic multiplier effect of ionization are generated around the anode, while

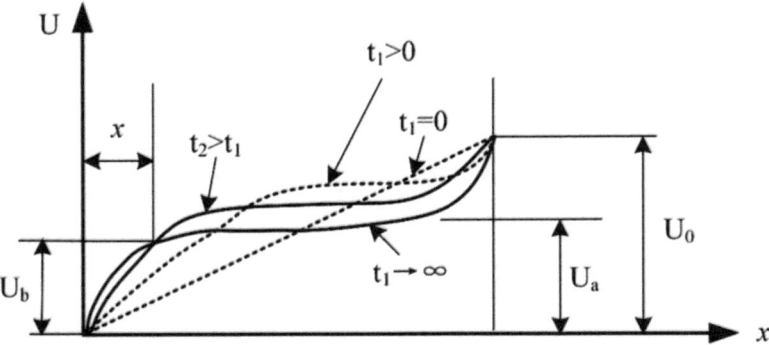

Fig. 3.6 The transient voltage in the corona charging (Hu 2008)

the electrons in plasma would shield on the charges of anode. The smooth part of the potential of the gas dielectric (in plasma state) first occurs close to the anode and then extends to the cathode very quickly. Thus, the potential of the anode decreases with U_a, while the potential of the cathode increases with U_b. There is little change of the potential of the gas dielectric between the anode and the cathode. Nearly all the voltage is applied in the region, $0 < x < d$, thus the spark charge comes into the stable gaseous discharge. The applied voltage between the two electrodes is $U = U_b$-$U_a < U_0$. In other words, the voltage between the two electrodes in the complete corona discharging is less than that of the induction charging. The electronic multiplier effect of ionization and the electrons motion alter the electric field simultaneously enough that energy is needed to maintain the kind of electron movement. At the initial voltage of spark discharge, U_0, the condition for the corona would be just created, and the charging process is extremely unstable. Sometimes it works in the induction charging, and sometimes it works in the corona charging. Only if the applied charging voltage is a little higher than the initial voltage, the anode could have enough energy to maintain the stable gaseous discharging, promoting the full corona charging.

The transient process in corona charging is characteristic of the sharp decrease of the voltage and the surge increase of the current. Only in a very short time, the corona charging ends its transient state from the induction charging, and enters into a steady corona charging. In experiment, only if the applied voltage is changed extremely slowly, the transient state changing from the induction charging to the corona charging could be achieved, characteristic of not steady charging and the surge increase of the current, described by the inverted Z shape change between the applied voltage and the current in charging process, as illustrated in Fig. 3.7.

For electrospray, theoretically, the induction charging could be employed in the charging process under the low applied voltage. At the stage of induction charging with the invariable external circuit parameters, the charged quantity (electric current) increases along with the applied voltage. With the increase of the applied electric field intensity, the molecules in the gas dielectric move faster, therefore the induction charging is confined by the applied charging voltage limit. Once the applied voltage reaches the startup voltage of the gas spark discharge, U_0, the

Fig. 3.7 The relationship of the applied voltage and the current in corona charging

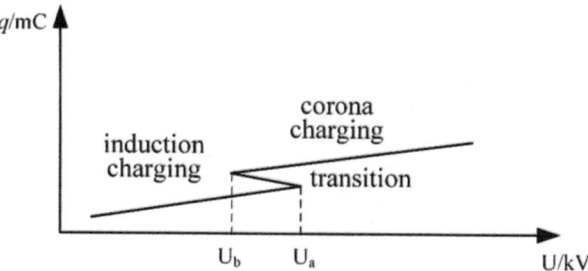

motion of molecular become intense enough to make the molecular ionized, and hence the corona discharge is achieved. Based on the electric circuit model of the corona charging, the transient process from the induction charging into the corona charging causes the sudden reduction of the resistance of the gas dielectric, thus resulting with the sharp reduction of the voltage between the electrodes but the surge increase of the current. Consequently, the charging process enters into the complete corona charging condition, and the charged quantity is thus determined by the applied voltage and the conductivity of the ionized gas dielectric between the two electrodes. The charged quantity would increase linearly with the applied voltage.

The goal of electrostatic spray is to gain a larger amount of charges on droplets. The optimum operation parameters are the key in electrospray technology.

3.2.4 Interface Charging

The electrification of thunder clouds are mainly centered within the mixed phase zone where all kinds of hydrometeors coexist, and the cause is basically thought to the result of the separation of OH^- ions from H^+ ions. Because the most abundant condensable substances in the solar system are water, hence we shall briefly regard the electrification of H_2O clouds as the main charge generation processes. So far, there have been two mainly driven mechanisms to explain the charge diffusion inside hydrometeors: the temperature gradient and the chemical potential gradient.

3.2.4.1 Temperature Gradient

The charge separation driven by the temperature gradient is based on the thermo-electric effect of ions. Since the dissociation of water is a function of temperature and H^+ ions have a higher mobility than OH^-, therefore more H^+ are accumulated in the colder regions of the particles. For example, when the environment is supersaturated with respect to the water, the water droplet will grow up and its water-vapour interface will acquire the negative charge due to the release of latent heat. However, if the environment becomes drier, the water droplet would

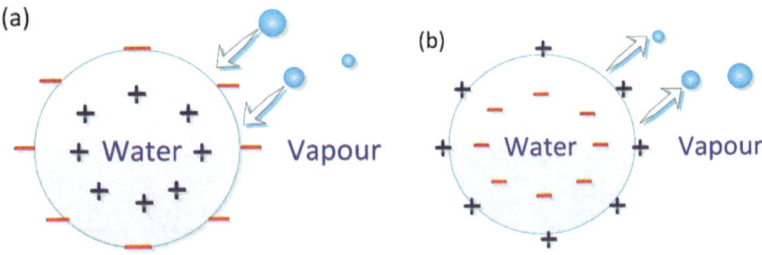

Fig. 3.8 Schematic of charge separation inside hydrometeor due to temperature gradient at different conditions. (**a**) When in a condensation state, the interface charge negative; (**b**) When in an evaporation state, the interface charge positive

evaporate, cooling the surface, and the charge will be reversed, as shown in Fig. 3.8. It also applies to ice-vapour interface.

3.2.4.2 Chemical Potential Gradient

The charge separation driven by the chemical potential gradient is based on the spontaneity of ions. For a water droplet in air, the density of water molecules which are located in the interfacial region is less than that of inside water.

The condition for chemical equilibrium is

$$\mu_{H_2O} = \mu_{H^+} + \mu_{OH^-}.$$

Then, the chemical potentials can be represented as:

$$\mu_{H^+} = \mu_{H^+}^0 + kT \ln C_{H^+},$$

and

$$\mu_{OH^-} = \mu_{OH^-}^0 + kT \ln C_{OH^-}.$$

It's assumed that the bulk water phase to be equilibrium with the interfacial region. Because of the reduced density, the dielectric constant of the interfacial region will be correspondingly reduced. The chemical potential of an ion in this region will, therefore, be increased by an amount known as the Born energy. The increase in the chemical potential leads the ions to transfer from the region 2 with a dielectric constant ε_2 to the region 1 with a larger dielectric ε_1, and the expression is given by

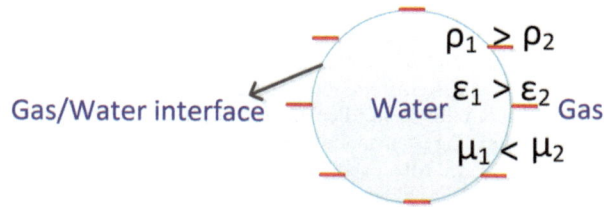

Fig. 3.9 Schematic of charge separation inside hydrometeor due to potential gradient. The interface charge negative

$$\Delta\mu(\text{Born}) = -\frac{(ze)^2}{8\pi\varepsilon_0 r}\left[\frac{1}{\varepsilon_2} - \frac{1}{\varepsilon_1}\right]$$

Since the Born energy is inversely proportional to the ionic radii, and the ionic radius of H^+ ions is less than that of OH^- ions, the Born energies of H^+ is larger than OH^-. So there exists an excess of negative charge within the interfacial region (Schechter et al. 1998), as shown in Fig. 3.9.

Then the chemical potential of H^+/OH^- in the interfacial region are expressed as:

$$\mu_{H^+}^{I} = \mu_{H^+}^{0} + kT \ln C_{H^+}^{I} + e\varphi + \Delta\mu_{H^+}(\text{Born})$$

and

$$\mu_{OH^-}^{I} = \mu_{OH^-}^{0} + kT \ln C_{OH^-}^{I} + e\varphi + \Delta\mu_{OH^-}(\text{Born}).$$

We can see that if the temperature increases, the chemical potential of H^+/OH^- will be enhanced, leading to more excess negative ions in the interfacial. So once the hydrometeor is not in thermal equilibrium, the surface charge on it will be continuously changed, even a big jump. In fact, when the hydrometeor is in thermal equilibrium, the net negative accumulated in the interface due to chemical potential gradient is very little, so when in non-thermal equilibrium, its change may be regard as an auxiliary of charge separation due to temperature gradient.

The temperature gradient and chemical potential gradient induce ion diffusion inside hydrometeors, and the difference of mobility of H^+ and OH^- produces a charge separation inside hydrometeors, resulting in different surface potentials for different size hydrometeors. Once they have a contact or collision, charges transfer between hydrometeors driven by surface tension gradient or potential gradient.

References

Chapman, S., and T.G. Cowling. 1958. *The mathematical theory of non-uniform gases: An account of the kinetic theory of viscosity, thermal conduction, and diffusion in gases.* Cambridge University Press: Cambridge.

Hu, B. 2008. *Study on the properties of corona plasma discharge atomization and design of sprayer heads(in Chinese)*. Beilin: Master, Chang'an University.

Jaworek, A., and A.T. Sobczyk. 2008. Electrospraying route to nanotechnology: An overview. *Journal of Electrostatics* 66 (3-4): 197–219.

Jung, J.H., S.Y. Park, J.E. Lee, C.W. Nho, B.U. Lee, and G.N. Bae. 2011. Electrohydrodynamic nano-spraying of ethanolic natural plant extracts. *Journal of Aerosol Science* 42 (10): 725–736.

Raizer, Y.P., J.E. Allen, and V.I. Kisin. 2011. *Gas discharge physics*. London: Springer.

Schechter, R.S., A. Graciaa, and J. Lachaise. 1998. The electrical state of a gas/water interface. *Journal of Colloid & Interface Science* 204 (2): 398–399.

Spitzer, L. 1962. *Physics of fully ionized gases*. New York/London: Interscience Publishers.

Wadhwa, C.L. 2007. *High voltage engineering*. New Delhi: New Age International (P) Limited, Publishers.

Chapter 4
Numerical Modeling Methods for Droplet Electrification

4.1 Governing Equations of Fluid Flow and Electric Field

4.1.1 Governing Equations of Fluid Flow

In electrohydrodynamics, inertia and viscosity of the fluid and the electric field force are coupled together, directly influencing the fluid motion. For incompressible fluid, equation of mass conservation and conservation-of-momentum equation are:

$$\nabla \cdot \mathbf{U} = 0 \tag{4.1}$$

$$\frac{\partial (\rho \mathbf{U})}{\partial t} + \nabla \cdot (\rho \mathbf{U}\mathbf{U}) = -\nabla p + \nabla \cdot \left(\sigma^{f} + \sigma^{e}\right) + \mathrm{f}_{b} \tag{4.2}$$

where:

$\mathbf{U}$ fluid velocity field, m·s^{-1};
p pressure, Pa;
ρ density, kg·m^{-3};
f_{b} volume force, N;

Viscous stress tensor σ^{f} can be calculated from the following equation:

$$\sigma^{f} = \mu\left[\nabla \mathbf{U} + (\nabla \mathbf{U})^{T}\right] - \frac{2}{3}\mu(\nabla \cdot \mathbf{U})\mathbf{I} \tag{4.3}$$

In which $\mathbf{I}$ represents unit tensor, μ represents coefficient of dynamic viscosity.

Maxwell stress tensor σ^{e} and its corresponding electric field force f^{e} can be calculated respectively in following equations:

© Springer Nature Singapore Pte Ltd. 2017
Z. Gu, W. Wei, *Electrification of Particulates in Industrial and Natural Multiphase flows*, DOI 10.1007/978-981-10-3026-0_4

$$\sigma^e = \varepsilon\varepsilon_0 \mathbf{E}\mathbf{E} - \frac{\varepsilon\varepsilon_0}{2}\mathbf{E}\cdot\mathbf{E}\left(1 - \frac{\rho}{\varepsilon\varepsilon_0}\frac{\partial\varepsilon\varepsilon_0}{\partial\rho}\right)\mathbf{I} \tag{4.4}$$

$$\mathbf{f}_e = \nabla\cdot\sigma^e = q_V\mathbf{E} - \frac{1}{2}E^2\nabla\varepsilon\varepsilon_0 + \nabla\left(\frac{1}{2}\rho\frac{\partial\varepsilon\varepsilon_0}{\partial\rho}E^2\right) \tag{4.5}$$

In which ε represents the dielectric coefficient of fluid, $\varepsilon_0 = 8.85 \times 10^{-12}$ $C\cdot V^{-1}\cdot m^{-1}$, the dielectric coefficient in vacuum, q_V represents the volume charge density of the interior of fluid, $\mathbf{E}$ represents the electric field. The first item to the right of the equation is caused by electric charges and electric field, alongside the direction of electric field. The second right item is the polarized stress of electric field, alongside the normal direction of interfaces of fluid and surrounding media. The third right item is the additional force caused by the changes in material density, which should be ignored since it is assumed at the beginning of this chapter that the fluid is incompressible.

Normally, dynamic current generated in the field of electrical hydromechanics is small; hence the magnetostriction effect can be ignored. Therefore, electric field $\mathbf{E}$ is considered to be irrotational field ($\nabla \times \mathbf{E} = 0$). Then Gauss Theorem of fluid with the relative dielectric coefficient as ε can be described through electrostatic displacement ($\mathbf{D} = \varepsilon\varepsilon_0\mathbf{E}$) as:

$$\nabla\cdot\mathbf{D} = q_V \tag{4.6}$$

In which q_V represents the volume charge density.

Electric field strength $\mathbf{E}$ can be calculated through the potential gradient ϕ:

$$\mathbf{E} = -\nabla\phi \tag{4.7}$$

The conservation equation of charges can be written as:

$$\frac{Dq_V}{Dt} + \nabla\cdot(K\mathbf{E}) = 0 \tag{4.8}$$

In which K represents the conductivity of fluid. $D/Dt = \partial/\partial t + \mathbf{U}\cdot\nabla$ is the material derivative.

Tangential components crossing electric field strength of interfaces are continuous, $\mathbf{n} \times \|\mathbf{E}\| = 0$, in which $\|\cdot\|$ represents variables changing from dielectric 1 to dielectric 2, and n represents the unit normal vector of the interfaces. However, normal components crossing electric field strength of interfaces are noncontinuous which can be expressed as:

$$\mathbf{n}\cdot\|\mathbf{D}\| = q_s \tag{4.9}$$

in which q_s represents the surface charge density.

4.1.2 Gas-Liquid Two-Phase Interface Tracking

In traditional VOF (Hirt and Nichols 1981), it simultaneously solves indicator function, continuity equation and momentum equation representing volume fraction of a certain phase:

$$\frac{\partial \gamma}{\partial t} + \nabla \cdot (\mathbf{U}\gamma) = 0 \tag{4.10}$$

In which γ represents the volume fraction (hereinafter referred to as VF); γ of a certain phase is in the range of $0 \leq \gamma \leq 1$, when the value equals 0 or 1, there is only one phase in the area, or it represents the interface as shown in Fig. 4.1.

When there are two kinds of immiscible fluid in an area, it is considered that only one of them is the main fluid ($\gamma = 1$), physical characteristics of the fluids in the area can be calculated by the weighted average of VOF; therefore, changes only happen on the interfaces:

$$\rho = \rho_l \gamma + \rho_g (1 - \gamma) \tag{4.11}$$

$$\mu = \mu_l \gamma + \mu_g (1 - \gamma) \tag{4.12}$$

$$K = K_l \gamma + K_g (1 - \gamma) \tag{4.13}$$

$$\varepsilon = \varepsilon_l \gamma + \varepsilon_g (1 - \gamma) \tag{4.14}$$

In which l and g respectively represents gaseous phase and aqueous phase.

A key issue of proceeding numerical simulation of free surface by VOF model is how to ensure the conservation of phase fraction. Especially when the density contrast of two phases is relatively large, minor VF error will cause major error

Fig. 4.1 Definition of volume fraction

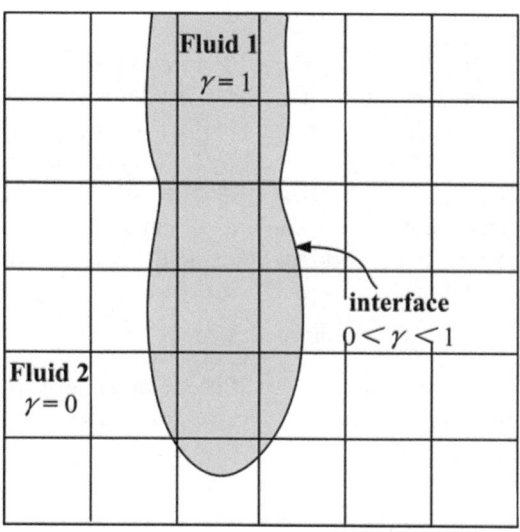

of physical characteristics during the calculation; hence it is particularly important to ensure the conservation of phase fraction.

Generally, due to the high sensitivity of grid resolution, multiple grids are required in the interface area between two phases. However, it is pretty difficult to ensure the boundedness and conservation of phase fraction at the same time. In order to overcome these problems, researchers have done a lot of effort, while in traditional VOF method; the two phases share an equally velocity.

A modified VOF method (Rusche 2002) is adopted. An additional convection term calculated from the weighted average of the corresponding velocity of fluid and gas is added to the transport equation of phase fraction to solve the more acute interface problems. In this approach, if a separate solution of phase fraction equations in each phase is required, then the phase fraction equations in each phase will be written as:

$$\frac{\partial \gamma}{\partial t} + \nabla \cdot (\mathbf{U}_l \gamma) = 0 \tag{4.15}$$

$$\frac{\partial (1 - \gamma)}{\partial t} + \nabla \cdot \left[\mathbf{U}_g (1 - \gamma) \right] = 0 \tag{4.16}$$

Assume the velocity of gaseous phase and aqueous phase is proportional to the effects of free surface evolution and the corresponding phase fraction, and define weighting expression of equivalent velocity as (Berberovicacute et al. 2009):

$$\mathbf{U} = \lambda \mathbf{U}_l + (1 - \gamma)\mathbf{U}_g \tag{4.17}$$

The equation can be rewritten as:

$$\frac{\partial \gamma}{\partial t} + \nabla \cdot (\mathbf{U}\gamma) + \nabla \cdot [\mathbf{U}_r \gamma (1 - \gamma)] = 0 \tag{4.18}$$

In which $\mathbf{U}_r = \mathbf{U}_l - \mathbf{U}_g$ represents relative velocity vector. Compared with the original equation, it includes an additional convection term, which will only act on the interfaces and have no effects on areas beyond the interface.

In VOF calculation, volume force f_b of the equation contains gravity and surface tension of the interfaces. Interfaces of gas-fluid two-phase will generate additional pressure gradient, which is presented as volume force and can be calculated from the continuum surface force model (Brackbill et al. 1992):

$$f_{st}^V = \sigma \kappa \nabla \gamma \tag{4.19}$$

In which σ represents the coefficient of surface tension, κ represents the curvature of interfaces which can be calculated from the following equation:

$$\kappa = -\nabla \cdot \left(\frac{\nabla \gamma}{|\nabla \gamma|} \right) \qquad (4.20)$$

4.2 Coupling with Charging Model

4.2.1 Electrical Leakage Model

According to the Maxwell Equation, characteristic time of electric field $\tau_C = \varepsilon\varepsilon_0/K$ can be considered as the ratio of dielectric coefficient to conductivity of fluid, and temporal scales of motion can be considered as viscosity relaxation time τ_P. For isotropous non-dielectric continuum fluid, the time of charge being redistributed and achieving a steady state is far less than that of fluid motion ($\tau_P >> \tau_C$), hence the changing situation of electric charges with time can be ignored and the equation can be simplified as (Saville 1997):

$$\nabla \cdot (K\mathbf{E}) = 0 \qquad (4.21)$$

Under the influence of external electric field, electric field stress generated in the fluid can be calculated from the Maxwell Stress Tensor (Equation). In this research, its equivalent volume force can be expressed as:

$$\mathbf{f}_e^V = q_V\mathbf{E} - \frac{1}{2}\mathbf{E}^2\nabla\varepsilon\varepsilon_0 \qquad (4.22)$$

Since the electric charges are distributed on the surfaces of liquid droplets, in eq. (4.22), the acting force is actually acting between surface charges and electric field, which is the force of area. Surface charge density of liquid droplets can be calculated through the difference value of electrically displaced normal components passing through the interfaces:

$$q_s = \mathbf{n} \cdot \|\mathbf{D}\| \qquad (4.23)$$

In which q_s represents charge density, $\|\mathbf{D}\|$ represents difference value of electric displacement in different fluids, and $\mathbf{n}$ represents the unit normal component of interfaces.

Taking the same approach of surface tension, electric force on the surfaces of liquid droplets can be transformed into volume force:

$$\mathbf{f}_e^V = q_s\mathbf{E}\nabla\gamma - \frac{1}{2}\mathbf{E}^2\nabla\varepsilon\varepsilon_0 \qquad (4.24)$$

4.2.2 Constant Charge Model

When connecting conductive liquid droplets with electrodes, net charges will be formed on the surfaces of liquid droplets. Under the influence of external electric field, electrified liquid droplets will perform different deformational behavior from that of neutral liquid droplets. Assuming free charges is well-distributed on the surfaces of liquid droplets, then the surface charge density of fluid droplets is constant, namely q_s in the Eq. (4.24) is constant. The rest of the solutions are similar to electrical leakage model.

4.3 Applications of Numerical Simulation on Droplet Electrification

4.3.1 Movement of Electrified Liquid Droplets in Electric Fields

In electrohydrodynamics, flow field and electric field are coupled with each other. Fluid motion is influenced by electric-field distribution and surface charge density, and vice versa. Fluid motion can change the surface charge density distribution of interfaces. With initial shape as sphere, neutral leaky dielectric droplet or electrified liquid droplets in the electric field are utilized to investigate the deformation and mechanical behavior of liquid droplets under the effect of electric field force.

Figure 4.2 illustrates the schematic of geometry model of droplet suspending in another immiscible liquid under external electric field. The droplet of density ρ_i, viscosity μ_i, permittivity ε_i, and electrical conductivity K_i, suspended in an immiscible fluid of density ρ_o, viscosity μ_o, permittivity ε_o, and electrical conductivity K_o. The droplet suspension consists of two immiscible phase, the inner phase represented by subscript i and the outer phase represented by subscript o. At the initial stage, the shape of the droplet is assumed to be spherical, after applying different electric potentials to the upper and lower electrodes, and steady electric field is generated. Under the influence of electric field, the droplet may start to deform depending on the operating conditions. The interface separating the two fluids is assumed to have a constant interfacial tension coefficient σ. The following fluid property ratios are defined: $R = K_o/K_i$, $Q = \varepsilon_i/\varepsilon_o$, $M = \rho_i/\rho_o$, $\lambda = \mu_i/\mu_o$. The potential difference applied to the upper and lower parallel electrodes to form an uniform electric field between the electrodes; in order to obtain non-uniform electric field, an inverted T-shape electrode is used, as shown in Figs. 4.2a and b.

Velocity boundary of upper and lower plates is no-slip boundary. Symmetry boundary is set respectively to the left and right side in order to eliminate the influence of left and right boundary on mechanical behaviors of liquid droplets. Detailed settings of boundary conditions are shown in Table 4.1. When electric field is uniform, the computational domain is chosen as 0.02 m × 0.02 m, diameter of

Fig. 4.2 Schematic of the geometry model of droplet suspending in another immiscible liquid under (**a**) an external uniform electric field and (**b**) an external non-uniform electric field

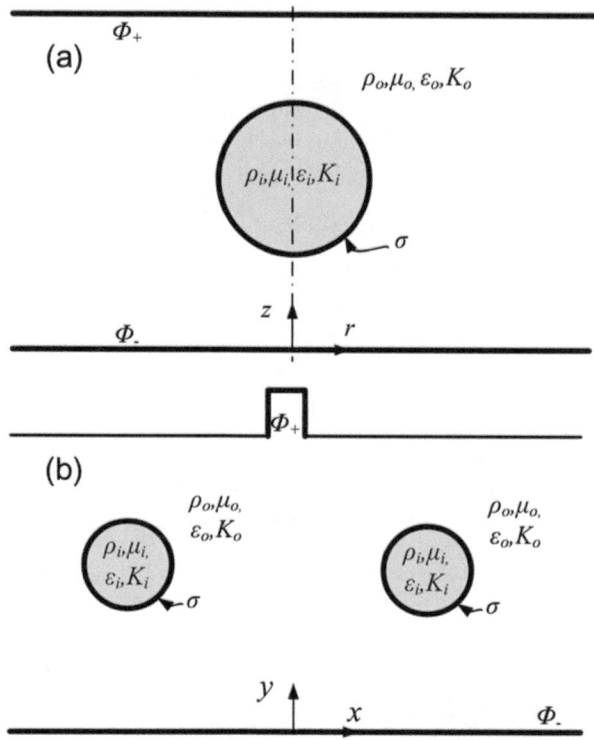

Table 4.1 Settings of boundary conditions

	Velocity	Pressure	Voltage
Left boundary	Symmetry boundary	Symmetry boundary	Zero gradient
Right boundary	Symmetry boundary	Symmetry boundary	Zero gradient
Upper polar plate	No slip	Zero gradient	Φ_0
Lower polar plate	No slip	Zero gradient	0

liquid droplets as 0.005 m, computational grid as 400 × 400, with local refinement applied to the vicinity of droplets. When electric field is non-uniform, the computational domain is chosen to be blade diameter of 0.05 m × 0.01 m, size of upper protuberance as 0.0015 m × 0.003 m, diameter of liquid droplets as 0.002 m, and computational grid as 800 × 200, with local refinement applied to the vicinity of droplets. Initially, spherical liquid droplets are suspending between polar plates with their initial velocity as 0.

Under the framework of OpenFOAM (OpenCFD 2011), an open source software of computational fluid dynamics (hereinafter referred to as CFD), the finite volume method (hereinafter referred to as FVM) of unstructured grids is adopted. Second order implicit algorithm are employed for unsteady terms, QUICK scheme for convection terms, and central difference scheme for diffusion terms, with the time step of 10^{-4} s. Firstly, by solving the equation of electric field equations, electric field distribution between upper and lower polar plates can be obtained, then the volume force of electric field acting on flow field can be calculated. Secondly, calculating the fluid flow government equation with VOF equation tracing the interface between liquid droplets and surrounding fluid, shape and position of liquid droplets after deformation/movement can be obtained.

① Deformation of neutral leaky dielectric droplet in uniform electric field

Define capillary number Ca_E to describe the relationship between electric field stress and surface tension:

$$Ca_E = \frac{a\varepsilon_o E_\infty^2}{\sigma} \tag{4.25}$$

In the equation, E_∞ represents the external electric field. Liquid droplets will deform under the effect of external electric field E_∞. As it is shown in Fig. 4.3, deformation coefficient of liquid droplets D is defined as follow:

$$D = \frac{L - B}{L + B} \tag{4.26}$$

In which L represents the amount of deformation of liquid droplets along the direction of electric field, B represents the amount of deformation of liquid droplets perpendicular to the direction of electric field. A positive D represents a deformation of the droplet that has an increased length in the axial direction (prolate), while a negative D represents a deformation of the droplet that has an increased length in the radial direction (oblate).

Fig. 4.3 Sketch of liquid droplet deformation

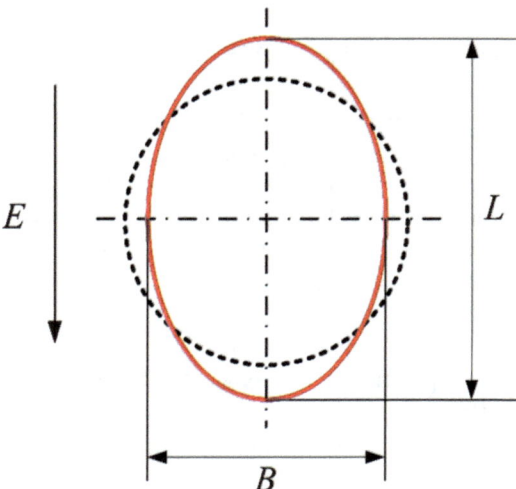

Assuming that liquid droplets are spherical and perform small deformation, Taylor (1966) and Ajayi (1978) respectively put forward the first order and the second order equation to calculate the deformation coefficient of liquid droplets D. Below is the expression of the second order equation.

$$D = k_1 Ca_E + k_2 Ca_E^2 \tag{4.27}$$

$$\left.\begin{aligned}
k_1 &= \frac{9}{16} \frac{F_d(R,Q,\lambda)}{(1+2R)^2} \\
k_2 &= \frac{k_1}{(1+2R)^2}\left[\left(\frac{9}{5}\frac{1-R}{1+2R} - \frac{1}{16}\right)F_d + R(1-RQ)\beta(\lambda)\right] \\
F_d(R,Q,\lambda) &= (1-R)^2 + R(1-RQ)\left[2 + \frac{3}{5}\frac{2+3\lambda}{1+\lambda}\right] \\
\beta(\lambda) &= \frac{23}{20} - \frac{139}{210}\frac{1-\lambda}{1+\lambda} - \frac{27}{700}\left(\frac{1-\lambda}{1+\lambda}\right)^2
\end{aligned}\right\} \tag{4.28}$$

In which $F_d(R,Q,\lambda)$ is known as Taylor's discriminating function, which is used to determines at first order the sign of D, i.e. it predicts whether the drop will deform into a prolate ($F_d > 0$) or an oblate ($F_d < 0$) shape. It depends on the physical properties of the droplet and the surrounding fluid medium.

In order to verify the proposed the numerical method, Fig. 4.4 gives deformation coefficient D obtained from simulation and deformation coefficient D calculated from the first order and the second order equation given different electric capillary number Ca_E. It can be inferred that with the increasing of external electric fields, deformation of liquid droplets is growing. When the deformation of liquid droplets is relatively small ($D < 0.12$), simulation results agree well with theoretical predictions. However, when the deformation of liquid droplets is relatively large, simulation results begin to deviate from theoretical values. Reasons for the

Fig. 4.4 Comparison between theoretical values and simulation values of relationship between deformation coefficient of liquid droplets and external electric fields. Calculation condition: $R = 0.2$, $Q = 2.0$, $\lambda = 1.0$

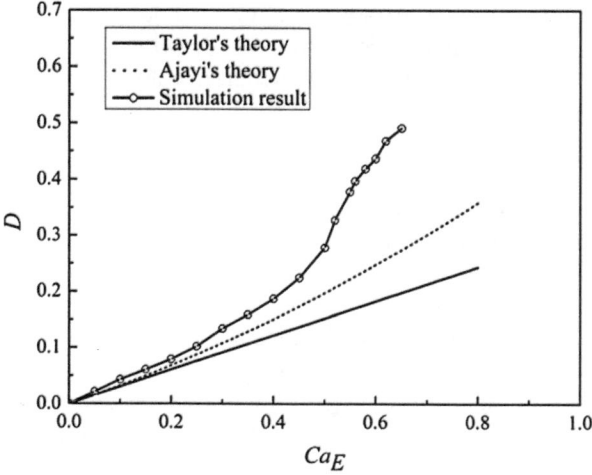

discrepancy may be that the theoretical formula is based on assumption of small deformation, which is not applicable to liquid droplets of large deformations.

The deformation of leaky dielectric droplet depends on the physical properties of the droplet and the surrounding fluid. Figure 4.5 shows two typical forms of deformation of neutral leaky dielectric droplet and their corresponding surface charge distribution and velocity field distribution of liquid droplets. When the relationship between physical property of liquid droplets and their surrounding fluid is $R = 2.0$, $Q = 2.0$, $\lambda = 1.0$, e.g. liquid droplets is silicon oil and their surrounding fluid is sextolphthalate, under the influence of electric field, free charges on spherical liquid droplets will be transferred and redistributed with the positive charges gathering on the top semisphere and negative charges on the bottom semisphere, As Fig. 4.5c shows. Under the upper-to-lower orientation electric field, the droplet would be deforms oblate, as shown in Fig. 4.5a.

As the ions are accumulated on the interface, a tangential force is generated on the interface and results in a steady circulating flow within the droplet. The fluid flows in a circulation pattern inside the droplet from the axis poles to the equator along the droplet surface, from the equator to the drop centers along the radial direction, and from the droplet center to the poles along the symmetric axis, as shown in Fig. 4.5e.

For droplets with the relationships of the fluid properties as: $R = 0.1$, $Q = 2.0$, $\lambda = 1.0$, e.g. droplet is castor oil, surrounding fluid is silicone oil. Under the external electric field force, the negative ions move to upper semisphere and the positive ions move to the lower semisphere(Fig. 4.5d), the droplet will be deforms prolate, as shown in Fig. 4.5b. At this moment, flow field inside the droplets is exactly opposite to that inside the prolate droplets. Fluid moves from the equator toward the poles along the surface of droplet, as well as moving from the droplet surface toward droplet center along the axial direction, as Fig. 4.5f shows. Under the effect of external electric field, recirculating motion generated inside the droplet can be the reference of the mixing method that fully dissolves the fluid inside droplets as well as protecting them from the pollution of external fluid.

It is worthy to note that, in the above-mentioned two kinds of electrofluid systems, liquid droplets are neutral, without net charges on their surfaces. Thus, under the influence of external electric field, the droplet only deforms and cannot moves.

② Deformation/motion of leaky dielectric droplet in Non-uniform electric field

When leaky dielectric droplet is in a non-uniform electric field, the electric force acting on the droplets' surfaces is non-uniform. In this case, droplets not only deform, but move along the direction of electric field lines. Due to the difference between the physical properties of droplets and external fluid, free charges on droplet surfaces are redistributed taking the normal direction of electric field lines as the symmetry axis under the effect of electric field force (as Fig. 4.6c,d shows), and free charges of liquid droplets accumulate at the sharp corner (as Fig. 4.6d shows). Eventually, as shown in Fig. 4.6a,b, it leads to the two different kinds of deformation/motion.

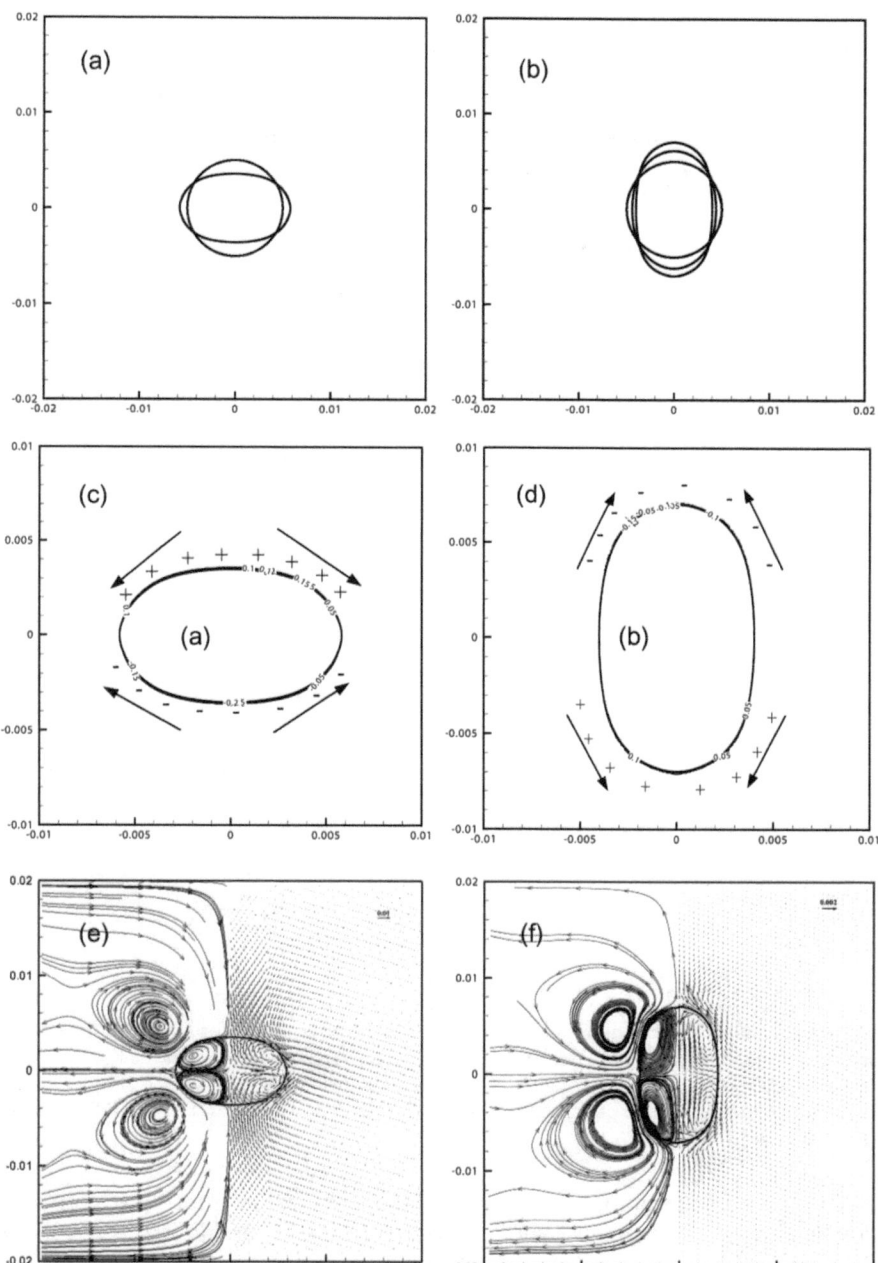

Fig. 4.5 Simulated deformation of leaky liquid droplet in an uniform electric field, in which the unit of coordinate is: m. Shape of liquid droplet: flat (**a**) and flat-long (**b**); distribution of free charges on droplet surface: flat (**c**) and flat-long (**d**); flow field surrounding the droplet: flat (**e**) and flat-long (**f**), in which speed lines are to the *left* and velocity vectors are to the *right*. Calculate condition: (**a**)(**c**)(**e**): $R = 2.0, Q = 2.0, \lambda = 1.0, Ca_E = 0.65$; (**b**)(**d**)(**f**): $R = 0.1, Q = 2.0, \lambda = 1.0, Ca_E = 0.5$

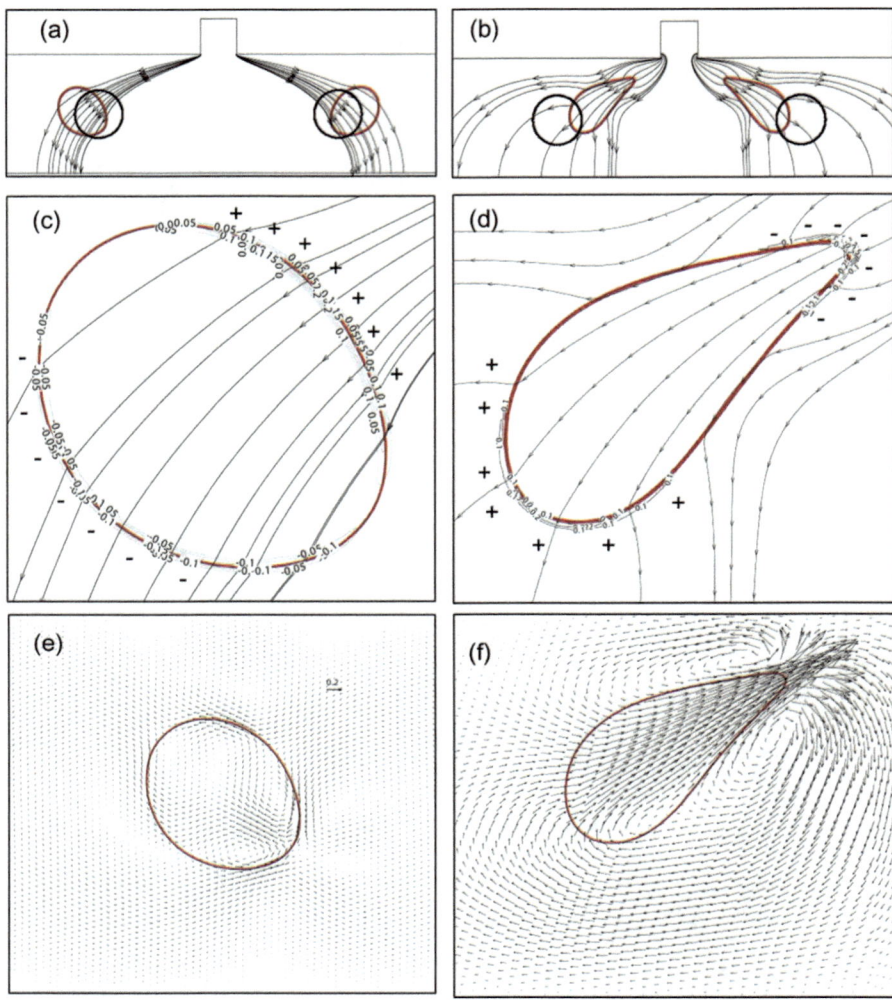

Fig. 4.6 Simulated deformation of leaky liquid droplet in a non-uniform electric field. (**a, b**) droplet deformation and electric field line distribution; (**c, d**) distribution of free charges on droplet surface (partial enlarged view); (**e, f**) velocity vector inside and outside the droplet (partial enlarged view). Calculation conditions: (**a, c, e**): $R = 2.0$, $Q = 2.0$, $\lambda = 1.0$; (**b, d, f**): $R = 0.1$, $Q = 2.0$, $\lambda = 1.0$

Due to the non-uniformity of electric field force acting on droplet surfaces, liquid droplets are no longer remaining at rest; meanwhile, tangential stress of electric field still makes internal and external fluid of droplets rotate. Distribution of velocity vectors of internal and external fluid of droplets are shown in Fig. 4.6e, f.

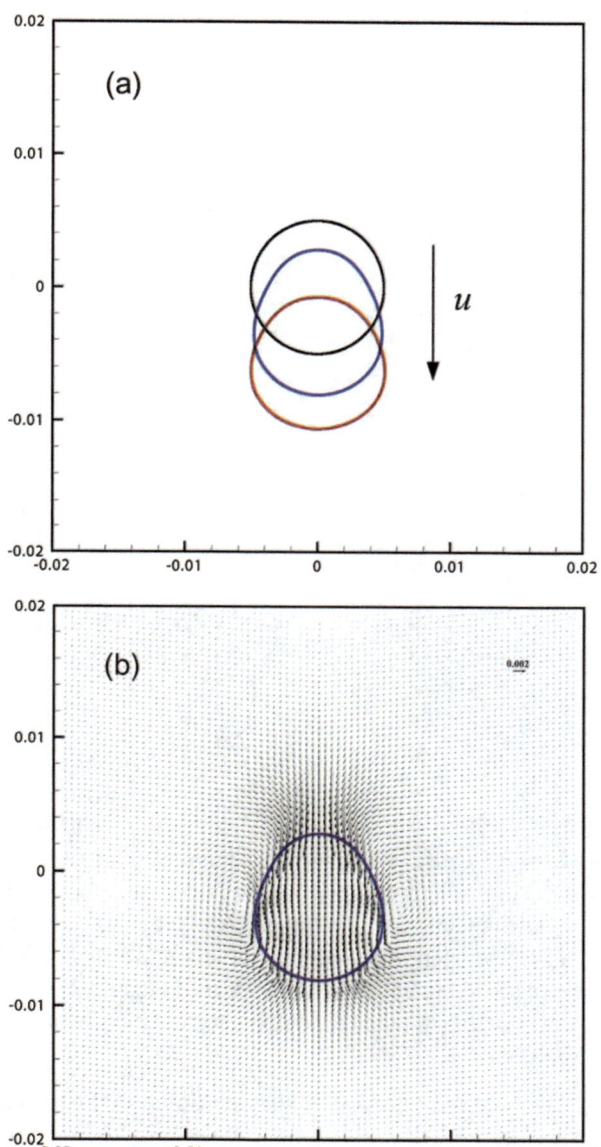

Fig. 4.7 Simulation of deformation and motion of charged liquid droplets in uniform electric field, in which the unit of coordinate is: m. (**a**) Status of droplet deformation and motion; (**b**) distribution of velocity vectors of internal and external flow field of charged droplets. Calculation condition: $R = 10.0$, $Q = 2.0$, $\lambda = 1.0$, $Ca_E = 1.0$, $q_s = 0.8$ C·m^{-2}

③ Deformation/motion of charged liquid droplets in an uniform electric field

Actually, if conductive droplets are exposed to polar plates or to high-voltage electric fields for a long time, droplet surfaces will be electrified and generate net charges. Under the effect of external electric field, charged droplets will deform and move. Figure 4.7 illustrates the relationship between properties of liquid droplets and external fluid $R = 10$, $Q = 2.0$, $\lambda = 1.0$, *e.g.* droplets is lubricant and the surrounding fluid is silicone oil, charged liquid droplets with surface charge density

Fig. 4.8 Simulation of
deformation and motion of
charged liquid droplets in
non-uniform electric field.
(**a**) Status of droplet
deformation and motion;
(**b**) distribution of velocity
vectors of internal and
external flow field of
droplets (partial enlarged
view). Calculation
conditions: $R = 10.0, Q = 2.0,$
$\lambda = 1.0, q_s = 0.8$ C·m^{-2}

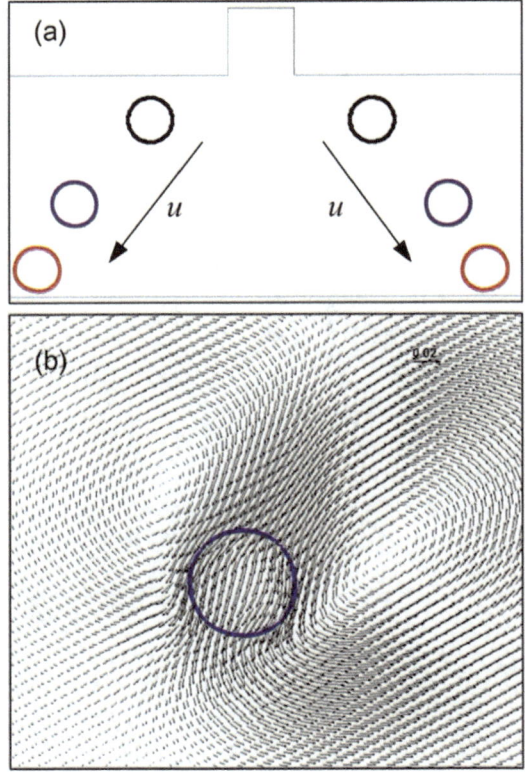

of $q_s = 0.8$ C·m^{-2} will deform and move under the effect of external electric field. It
can be inferred from Fig. 4.7a that under the effect of external electric field, charged
droplets will deform into the prolate shape and move along the direction of electric
field lines. Motion of liquid droplets drive external fluid downwards, and make
external fluid rotate under the effect of electric field tangential stress. However, as
shown in Fig. 4.7b, circulating vortex motion does not exist inside the liquid
droplets.

④ Deformation/motion of charged liquid droplets in a non-uniform electric field

Similar to what happens in uniform electric field, charged liquid droplets also
deform in non-uniform electric field. However, motion of droplets is dominated by
the Coulomb force, moving along the direction of electric field, as shown in
Fig. 4.8. Under such circumstances, if the strength of external electric field or
electric charge density of droplets is relatively high, macroscopic motion of droplets holds a leading role. Therefore, when analyzing such problems, liquid droplets
can be treated as solid particles which deformation can be ignored, only consider
the influence of electric field force on droplet trajectories.

4.3.2 Characteristic of Electrified Liquid in Electrostatic Spraying

Electrostatic spraying is a kind of high-precision technology for nano-devices fabrication. It is necessary to thoroughly analyze the entire process of electrostatic spraying for more precise control. The process is a multi-phase and multi-physical problem where flow fields and electric fields are coupled with each other. Besides, during the process, cone-jet mode will release charged liquid droplets. These charged droplets will generate self-induced electric field that changes the distribution of external electric field, which will affect the shape of cone-jet mode and the diameter of charged liquid droplets; thus the whole process becomes more complicated. The equivalent space-charge model of self-induced electric field generated by space charged droplets was proposed and takes the influence of self-induced electric field on surrounding electric field distribution of con-jet mode into consideration (Wei et al. 2013). It is subjected to the physical properties, flow rate of fluid and electrode characteristics (potential difference and height between the two polar plates and nozzle size), and hence is able to accurately the electric field distribution between nozzle and polar plates. This process allows the author to recur the formation and evolution of cone-jet mode, therefore provides the foundation to figure out the process of droplets being separated from the jet flow and the influence of applied voltage and flow rate on the formation process of cone-jet mode. Ultimately, it is thus made possible to predict the surface charge distribution of cone-jet mode under the condition of no operation and the current value generated by cone-jet mode.

The fluid with conductivity as K and dielectric coefficient as ε discharges from the nozzle. The diameter of nozzle as $d_{capillary}$ and height as L. The potential (voltage) of Φ_0 exerting on the nozzle and forming an external electric field pointing to the ground from the nozzle. The distance between nozzle outlet and the ground is $(H - L)$. Fluid flow Q is the input parameter of model. Physical quantities relating to the model are coefficient of surface tension σ, fluid viscosity μ and fluid density ρ between gas-liquid two-phase. Figure 4.9 shows the acting force considered in the calculation.

Most of the previous studies firstly assume the initial shape of the cone being Taylor Cone (Hartman et al. 1999) or the charge density distribution (Lim et al. 2011) on the surface of Taylor Cone, then calculate the formation and other relating parameters of cone-jet mode. Since the shape of the cone differs as operational parameter change, it is not always standard Taylor Cone. Besides, the shape of the cone would affect the size of jet flow it generated. Proceeding to the next step, it would also affect the current value generated in cone-jet mode. This study would not give assumptions on initial shape of liquid surface and charge distribution, namely the initial liquid surface is flat (along the nozzle outlet). On the contrary, this study analyzes the formation and evolution of cone-jet mode under the effect of external electric field from the temporal and spatial point of view, researches the

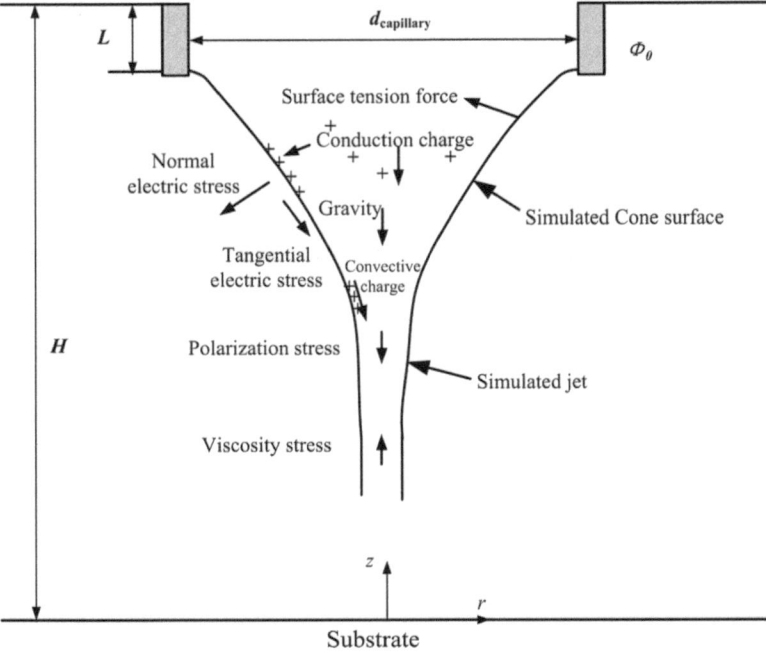

Fig. 4.9 Sketch of charge-transfer and fluid stress analysis in cone-jet mode

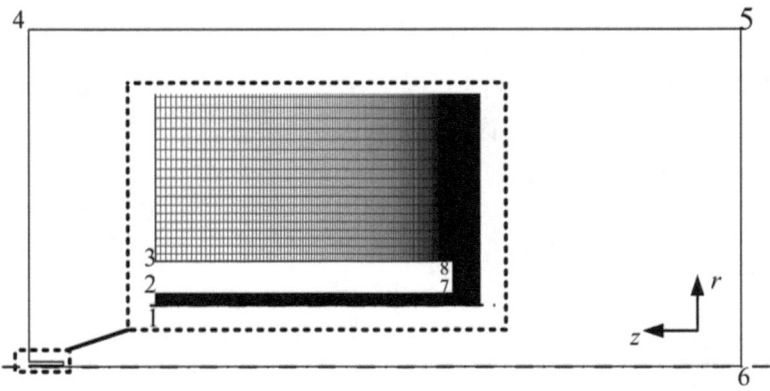

Fig. 4.10 Computational geometric domain and meshing

operational parameters affecting the shape of cone-jet mode, and forecasts the current value generated in cone-jet mode.

Figure 4.10 shows the selected computational domain; nozzle inner diameter $d_{\text{capillary}} = 0.12$ mm, external diameter of the nozzle is 0.45 mm, height of the nozzle $L = 1.5$ mm, distance from the nozzle to lower plate $H = 30$ mm. Total amount of computational grid is 408,500; gradient mesh with local refinement is adopted adjacent to the nozzle, minimum grid size is 2 μm.

Settings of boundary conditions and initial conditions are as follow:

Boundary 1–2 (velocity-inlet boundary): volume fraction of liquid $\gamma = 1$; circumferential velocity component $u_r = 0$; axial velocity component $u_z = -4Q/(\pi d^2_{capillary})$.

Boundary 2–7, 7–8 and 8–3: no-slip velocity boundary, $u_r = u_z = 0$; voltage boundary $\phi = \phi_0$ (ϕ_0 represents the applied external voltage).

Boundary 3–4: atmospheric boundary.

Boundary 5–6 (grounding electrodes): no-slip velocity boundary $u_r = u_z = 0$; voltage boundary $\phi = 0$.

Boundary 1–6: axisymmetric boundary.

At time t = 0, settings of initial conditions are: initial velocity of solution $u_z = -4Q/(\pi d^2_{capillary})$ and initial voltage on nozzle $\phi = \phi_0$.

By solving the electric field equations, the electric field distribution between the nozzle and the lower plate was obtained firstly. Next, the shape of the cone was simulated by using equations to calculate the body force of electric field acting on flow field. Then, by solving motion equations of fluid and taking VOF method, the tracking of gas-liquid interface is achieved. Finally, the formation and evolution of cone-jet can be obtained.

Shape of cone-jet mode depends on flow rate of fluid, voltage between polar plates and physical properties of fluid such as density, viscosity, conductivity, dielectric coefficient and coefficient of surface tension. Taking the experiment in literature (Tang and Gomez 1994) as reference, the study calculates the formation and evolution of cone-jet mode with heptane as the solution. Properties of the solution are illustrated in Table 4.2. This solution possesses characteristics of low conductivity and low dielectric coefficient. Examples in the article are calculated with applied voltage ranging from 3 kV to 8 kV and flow rate ranging from 1 cc·h^{-1} to 10 cc·h^{-1} to examine the formation and evolution of cone-jet mode under different voltages and flow rate, and forecast the quantity of current flow generated on cone-jet mode.

① Method validation

According to the experimental data, Tang (1994) has obtained the expression to calculate the diameter of heptane droplets separated from the tip of cone-jet flow zone: $d = 10.59 \times Q^{0.62}$. Figure 4.11 shows the comparison between simulated droplet diameters separated from the tip of cone-jet flow zone and droplet diameters obtained from experiments.

Table 4.2 Properties of solution

Density ρ(kg·m^{-3})	Coefficient of surface tension σ(N·m^{-1})	Dielectric coefficient ε	Conductivity K (S·m^{-1})	Viscosity μ(Pa·s)
684	0.018	1.9	1.15×10^{-6}	6.10×10^{-4}

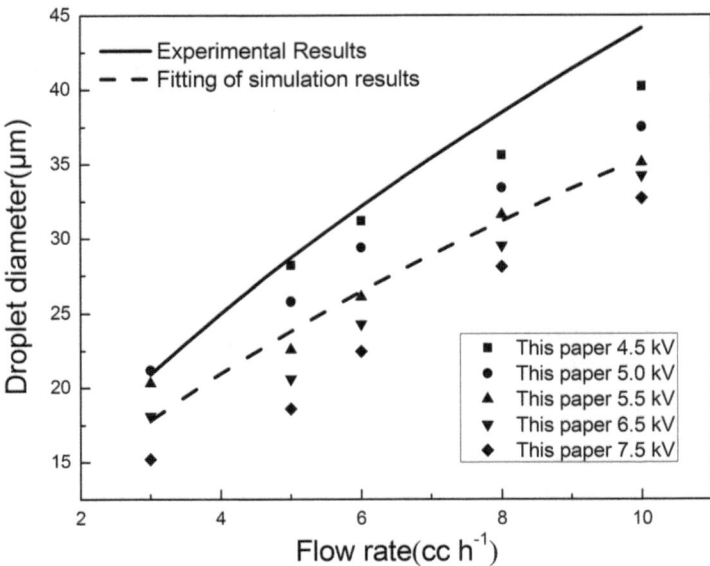

Fig. 4.11 Comparison between calculated droplet diameters and experimental values (Tang 1994)

Based on the simulation results, the relationship between diameters of separated droplets and flow rate can be expressed as: $d = 9.53 \times Q^{0.57}$. It can be inferred from Fig. 4.11 that when the flow rate is low, the simulated droplet diameters match well with the experimental values. As flow rate increases, the maximum error between simulated values and experimental values would reach to 20%, which indicates that when flow rate becomes larger, jet flow region, diameters, and velocity will also become larger, exacerbating the instability of jet flow and eventually leading to the error between simulated values and experimental values.

② Formation and evolution of cone-jet mode

There exists a potential difference between the nozzle and the cone tip. Under the joint effect of electric field of polar plates, gravity and surface tension, solution inside the nozzle recirculating in the internal nozzle, and free charges in the solution transfer from the fluid interior to fluid surface, which eventually electrifies fluid surface. Due to the normal discontinuity of interface dielectric displacement, fluid interface suffers the normal electric field force, which makes the interface move along the normal direction and leads to the contraction of it and the surrounding fluid moves along the tangential direction of the interface, thus forming jet flow whose diameter is smaller than that of the nozzle. With the continuously accelerated motion downwards of the jet flow, the interface shows its instability and eventually separates liquid droplets. As shown in Fig. 4.12, before separating the droplets, due to the effect of surface tension, the posterior interface of jet flow contracts and generates unstable structures looking like "neck". Under the effect of electric field force, jet flow is continuously lengthened. With the "neck" being thinner and

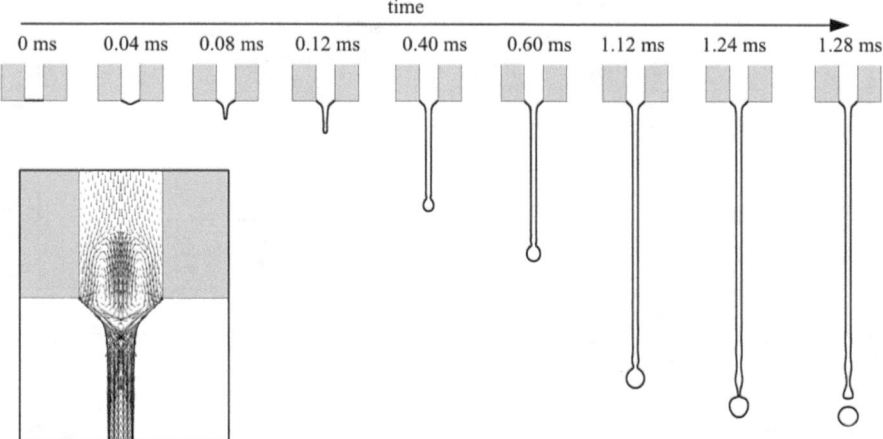

Fig. 4.12 Evolution of cone-jet mode

thinner, it is eventually pulled off, forming liquid droplets. It can be inferred that when cone-jet mode is stable, positions of separated droplets are roughly at the same height, and the length of jet flow is broadly unchanged.

Figure 4.13 shows the changes in velocity and electric field on the surface of cone-jet mode after forming a stable cone-jet mode, where the abscissa represents the length of cone-jet mode and the solid line shows the shape of cone-jet mode. The abscissa equal to 0 represents nozzle outlet while the maximum of abscissa represents jet flow tip. It can be inferred that at the jet flow tip, a liquid droplet with a diameter larger than that of the jet flow is generated. Since the nozzle is of certain thickness, electric field concentration is generated at the nozzle outlet; consequently, electric field generated by the surface of cone-jet flow near the nozzle outlet is of large magnitude. As being further away from the nozzle, concentration effect of electric field gradually diminishes; however, at the top of the cone, electric field mutates due to the discontinuous structure of cone-jet mode. After forming stable jet flow, electric field on the surface of it smoothly diminishes, as shown in Fig. 4.13a. Figure 4.13b is the variation diagram of velocity on the surface of con-jet mode. The maximum velocity appears at the inflection point of cone and jet flow, while velocity on the surface of jet flow shows the tendency of smooth and steady.

③ Initial voltage That Can form cone-Jet mode

Low voltage leads to small electric field strength while viscous stress and surface tension predominates, and at this moment, spraying mode would be dripping mode. With voltage gradually increasing, and electric field strength growing, electric stress predominates and the spraying mode would be cone-jet mode. Figure 4.14 shows the minimum voltage that can form cone-jet mode under different flow rate, in which the area below the curve represents dripping mode, while the area above the point on the curve represents cone-jet mode.

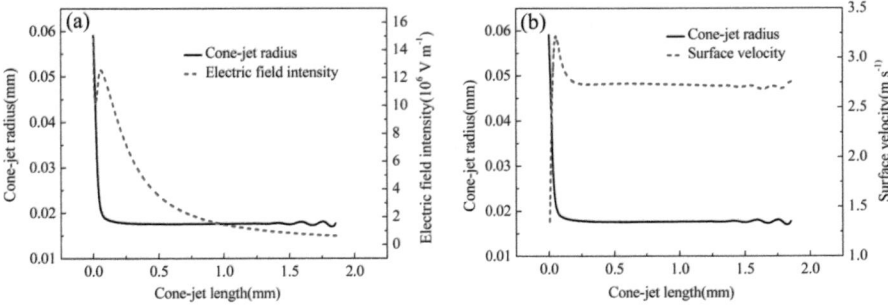

Fig. 4.13 (a) Distribution of electric field strength on the cone-jet surface; (b) distribution of velocity on the cone-jet surface

Fig. 4.14 Initial voltage of cone-jet mode

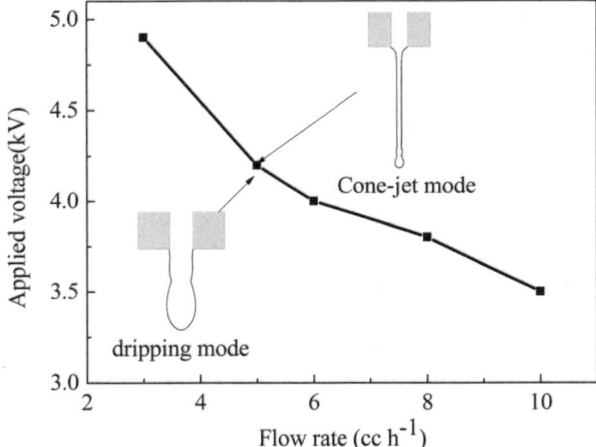

Fluid moves downwards driven by pressure, gravity and the tangential stress of electric field shown in Fig. 4.9; when flow rate increases, initial flow velocity of the fluid increases, making the tangential stress of electric field that drives the fluid moving downward correspondingly decreases. Thus the initial voltage that can form cone-jet mode decreases as flow rate increases as Fig. 4.14 shows, which agrees with the tendency discovered in the experiment in literature (Hayati et al. 1987).

④ Effects of voltage on the shape of cone-jet mode

As Fig. 4.15 indicates, with a fixed flow rate, if the voltage gradually increases, the cone angle will increase, the height of the cone will decrease and the diameter of jet flow would also decrease. In the literature (Tang and Gomez 1994), it is considered that under the condition of 10 cc·h^{-1} flow rate and 5.0 kV voltage, stable structure of cone-jet mode can be generated as well as liquid droplets of monodispersity, which matches well with calculations of the article.

Fig. 4.15 Effects of voltage on the shape of cone-jet flow. Flow rate: 10 cc·h^{-1}

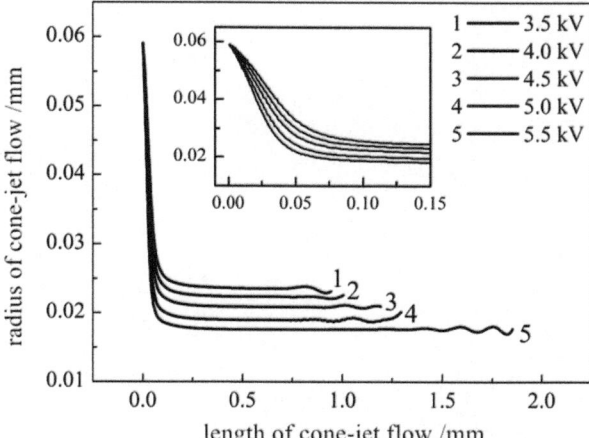

Fig. 4.16 Effects of flow rate on the shape of cone-jet flow. Applied voltage: 5.5 kV

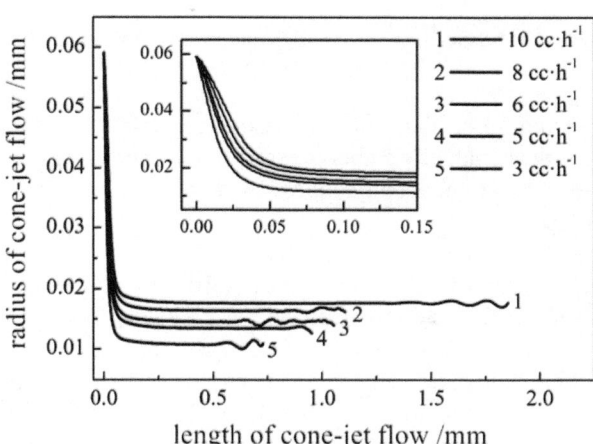

It can be inferred from Fig. 4.15 that as the voltage increases, length of cone-jet flow increases while instability of cone tip grows, forming multiple unstable structures shaped as "neck".

⑤ Effects of flow rate on the shape of cone-jet mode

It can be inferred from Fig. 4.16 that fixing the applied voltage, as flow rate decreases, the angle of the cone increases, the height of the cone decreases, and the diameter of jet flow decreases. When the flow rate is low, the length of jet flow is short, which makes it easier to separate liquid droplets.

⑥ Separation process of liquid droplets

The process of liquid droplets being separated from electrified cone-jet flow is expressed in Fig. 4.17. Due to the non-uniformity stress, the tip of cone-jet flow

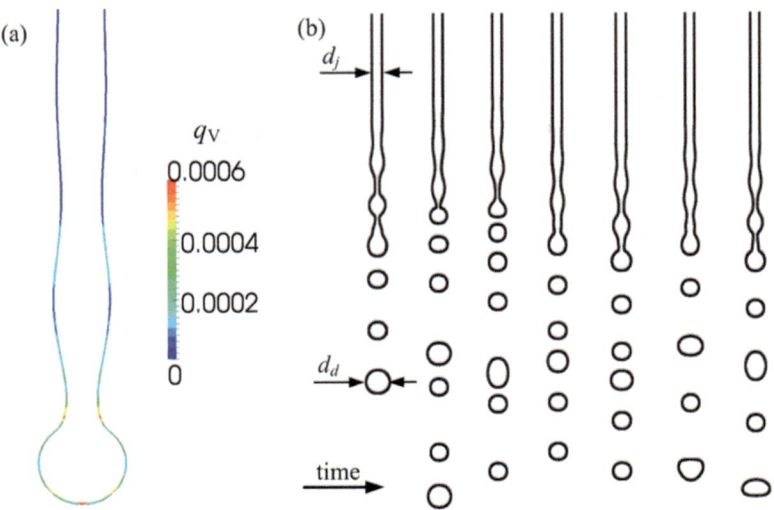

Fig. 4.17 Separation process of liquid droplets from cone-jet flow

becomes unstable, forming a corrugated shape. As the interface deforms, surface charges gradually gather together to the "neck" region in the corrugated area. These regions are gradually tapered off, and eventually pulled off, separating liquid droplets.

According to the classical Rayleigh Instability Theory: Without any electric charges, for fluid of low viscosity, the ratio of diameter of liquid droplets separated from jet flow d_d to diameter of jet flow d_j is 1.89.

The relationship between droplet diameter d_d and jet-flow diameter d_j, is obtained, which is shown in Fig. 4.18 where the slope of the line is 1.89, namely the ratio of d_d to d_j obtained from Rayleigh's instability. It can be inferred from Fig. 4.18 that d_d/d_j calculated is very close to 1.89, especially when the diameter is small, namely under the circumstance of low flow rate. This matches well with experimental results in the literature (Mutoh et al. 1979): value of d_d/d_j.

⑦ Current on the surface of cone-jet flow

Through experimental analysis and dimensional analysis, Fernández and Loscertales (1994) obtained the empirical formula for calculating the current quantity generated by cone-jet mode of fluids with high conductivity:

$$I = f(\varepsilon)(\sigma QK/\varepsilon)^{1/2} \tag{4.29}$$

In which:

$f(\varepsilon)$ factors relevant to dielectric coefficient.

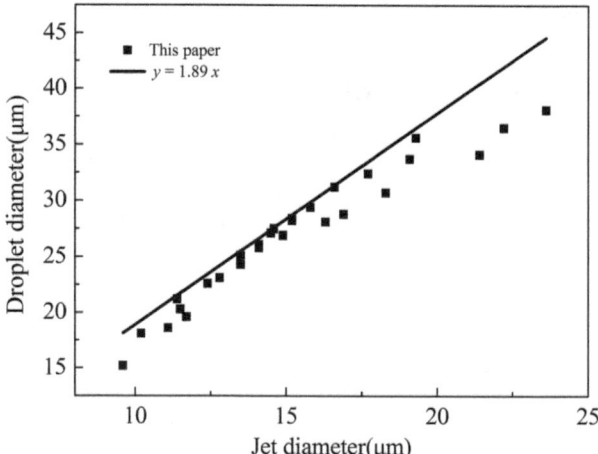

Fig. 4.18 Relationship between the diameters of jet flow and separated droplets

Similarly, according to the result of theoretical analysis, Ganan-Calvo (1997) put forward the scaling theory of calculating the droplet diameter:

$$I = 4.25 \left[QK\sigma / \ln \left(\frac{Q}{Q_0} \right)^{1/2} \right]^{1/2} \tag{4.30}$$

Ganan-Calvo (1999) also concluded the scaling theory of calculating the droplet diameter based on experimental results:

$$I = 2.6 \times I_0 (Q/Q_0)^{1/2} \tag{4.31}$$

In which:

$I_0 = \varepsilon_0^{1/2} \sigma \rho^{-1/2}$;
$Q_0 = \sigma \varepsilon_0 / (\rho K)$.

From the scaling theory it can be inferred that current generated in cone-jet mode is relevant to fluid conductivity K and flow rate Q instead of the applied voltage.

Using the method proposed in the chapter, current generated in cone-jet mode can be calculated as demonstrated in Fig. 4.19. The scaling theory proposed by Fernández and Loscertales is effective under the condition that the fluid is of high conductivity and flow rate is slightly higher than the minimum value, namely d_j (diameter of jet flow) divided by d_n (diameter of cone bottom) $d_j/d_n \rightarrow 0$. At the moment, it can be approximately considered that current generated in cone-jet mode is the current on the surface of the cone. However, when the fluid with low conductivity and flow rate is high, $d_j/d_n \approx 0.1$, current generated in cone-jet mode is a result of the concerted action of charge movement on cone surface and jet flow surface. Therefore, when using the scaling theory proposed by Fernández and

Fig. 4.19 Comparison between current values calculate by the article and the scaling theory

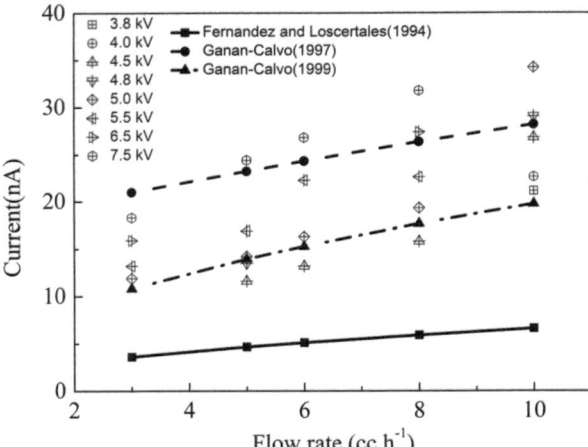

Fig. 4.20 Relationship between current and applied voltage under different flow rate

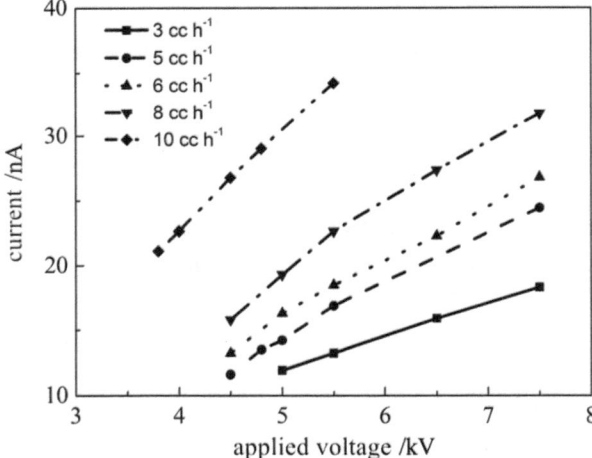

Loscertales to calculate current generated on the cone-jet flow surface of fluid of low conductivity, the result will be smaller.

Current value calculated by the method proposed in this chapter is roughly in line with that calculated by the scaling theory proposed by Ganan-Calvo (1997) and Ganan-Calvo (1999), as shown in Fig. 4.19. However, the former one can fluctuate significantly as voltage changes.

Figure 4.20 indicates the relationship between the current generated on cone-jet surface and the applied voltage. For fluids with low conductivity, current generated in cone-jet mode increases as applied voltage increases and flow rate increases. When flow rate reaches a high status, current and applied voltage is in a linear relationship since the diameter of jet flow formed at this moment is large; as flow rate decreases, the linear relationship between current and applied voltage weakens.

It is worth mentioning that when fluid with low conductivity is operated under large flow rate, the diameter of jet is thick; some controlling parameters that play almost no role at high conductivity or at small flow rate would affect the current, and the current varied almost linearly with the applied voltage.

References

Ajayi, O.O. 1978. A note on Taylor's electrohydrodynamic theory. *Proceedings of The Royal Society of London Series A-mathematical Physical A* 364 (1719): 499–507.

Berberovicacute, E., N.P. van Hinsberg, S. Jakirlicacute, I.V. Roisman, and C. Tropea. 2009. Drop impact onto a liquid layer of finite thickness: Dynamics of the cavity evolution. *Physical Review E* 79 (3): 036306.

Brackbill, J.U., D.B. Kothe, and C. Zemach. 1992. A continuum method for modeling surface tension. *Journal of Computational Physics* 100 (2): 335–354.

Fernández de la Mora, J., and I.G. Loscertales. 1994. The current emitted by highly conducting Taylor cones. *Journal of Fluid Mechanics* 260: 155–184.

Ganan-Calvo, A.M. 1997. Cone-jet analytical extension of Taylor's electrostatic solution and the asymptotic universal scaling laws in electrospraying. *Physical Review Letters* 79 (2): 217–220.

Gañán-Calvo, A.M. 1999. The surface charge in electrospraying: Its nature and its universal scaling laws. *Journal of Aerosol Science* 30 (7): 863–872.

Hartman, R.P.A., J.P. Borra, D.J. Brunner, J.C.M. Marijnissen, and B. Scarlett. 1999. The evolution of electrohydrodynamic sprays produced in the cone-jet mode, a physical model. *Journal of Electrostatics* 47 (3): 143–170.

Hayati, I., A.I. Bailey, and T.F. Tadros. 1987. Investigations into the mechanisms of electrohydrodynamic spraying of liquids: I. Effect of electric field and the environment on pendant drops and factors affecting the formation of stable jets and atomization. *Journal of Colloid and Interface Science* 117 (1): 205–221.

Hirt, C.W., and B.D. Nichols. 1981. Volume of fluid(VOF) method for the dynamics of free boundaries. *Journal of Computational Physics* 39 (1): 201–225.

Lim, L.K., J.S. Hua, C.H. Wang, and K.A. Smith. 2011. Numerical simulation of cone-jet formation in electrohydrodynamic atomization. *Aiche Journal* 57 (1): 57–78.

Mutoh, M., S. Kaieda, and K. Kamimura. 1979. Convergence and disintegration of liquid jets induced by an electrostatic field. *Journal of Applied Physics* 50 (5): 3174–3179.

OpenCFD, L. 2011. OpenFOAM: The OpenSource Computational Fluid Dynamics Toolbox (www.openfoam.com).

Rusche, H. 2002. *Computational fluid dynamics of dispersed two-phase flows at High phase fractions*. Ph.D. Thesis, Imperial College of Science, Technology and Medicine.

Saville, D.A. 1997. Electrohydrodynamics: The Taylor-Melcher leaky dielectric model. *Annual Review of Fluid Mechanics* 29: 27–64.

Tang, K. 1994. *The electrospray: Fundamentals and feasibility of its application to targeted drug delivery by inhalation*. Ph.D. thesis, Yale University.

Tang, K., and A. Gomez. 1994. On the structure of an electrostatic spray of monodisperse droplets. *Physics of Fluids* 6 (7): 2317–2332.

Taylor, G. 1966. Studies in electrohydrodynamics. I. Circulation produced in a drop by electric field. *Proceedings of The Royal Society of London Series A-mathematical Physical A* 291 (1425): 159–166.

Wei, W., Z. Gu, S. Wang, Y. Zhang, K. Lei, and K. Kase. 2013. Numerical simulation of the cone–jet formation and current generation in electrostatic spray—modeling as regards space charged droplet effect. *Journal of Micromechanics and Microengineering* 23 (1): 015004.

Part III
Basic Theory of Particle Charging in Multiphase Flows

Chapter 5
Charging Ways and Basic Theories of Particle Electrification

5.1 Charging Ways of Particle Electrification

5.1.1 Crack Electrification

After objects being cracked, there will be an imbalance of positive and negative charge distribution, which will generate static electricity known as crack electrification. In addition to the crack process that causes friction, sometimes charge distribution is already imbalanced before objects crack, which will also cause crack electrification. The quantity of crack electrification is relevant to the number and size of cleavage block, cracking velocity, unevenness of charge distribution before cracking, *etc*. Static electricity caused by crack electrification generally happened simultaneously to both sides of positively and negatively charged particles. Static electricity generated when solid is smashed is caused exactly by such reason.

5.1.2 Induced Electrification

Induced electrification usually happens to conductive particles. Due to static induction, objects being in an electrostatic field relocate the electric charges of conductors; thereby change the electric potential of objects. For insulating materials, being polarized in electrostatic field can also be electrified, which is also called induced electrification. Electric field of polarized insulating materials adsorbs certain free charges from the surrounding media toward itself to neutralize with bound charges carrying the opposite sign of insulating materials. After withdrawing the external electric field, these two charges of insulating materials are unable to restore electric neutrality. Therefore, the insulating materials will carry certain amount of electric charges.

© Springer Nature Singapore Pte Ltd. 2017
Z. Gu, W. Wei, *Electrification of Particulates in Industrial and Natural Multiphase flows*,
DOI 10.1007/978-981-10-3026-0_5

Similar phenomenon also happens in workplaces of powder industry. External electric field attracts powder particulates onto the conductors (such as metal equipment) of workplace. After neutralized charges of one polar to conductors, powder particulates will leave the conductors and be electrified. Induced electrification accounts for an additional electrification in the powder producing process.

After the action of strong electric field, some kinds of insulating materials, such as natural wax, resin, rosin, calcium titanate ceramics, polytetrafluoroethylene (teflon) and polypropylene (sulan), would remain polarized permanently or for a long time, forming electrets. Relaxation time of practical electrets lasts $3 \sim 10$ years.

5.1.3 Attached Electrification

The molecules of most matter are polar molecules, namely possessing dipole moment with dipoles aligned in orientation in the interface. On the other hand, particles carrying positive or negative charges are always floating in the air due to the effect of space electric field, various discharging phenomena, and cosmic rays, *etc.*. When these floating charged particles are adsorbed by dipoles on the surface of an object, or are attached to the object, the entire object will be electrified since it carries excess electric charges of a particular sign. If negative charges of dipoles aligned in orientation on the surface of the object are on the side of air, then the object surface will adsorb particles carrying positive charges in the air to make the object electrified. On the contrary, if positive charges of dipoles aligned in orientation on the surface of the object are on the side of air, then the object surface will adsorb particles carrying negative charges in the air to make the object negatively electrified. Quantity of attached electrification depends on factors such as the magnitude of dipole moment of molecules, arrangement of dipoles, cleanliness of the object surface and type of the air-floating charged particles.

5.1.4 Contact Electrification

Solid can be divided into metal conductor, insulator and semi-conductor. Contact of two different kinds of solid could be two dissimilar metals, two insulators or two semi-conductors. Also, it could happen between metals and insulators, metals and semi-conductors, or semi-conductors and insulators. Of all the theories concerning contact electrification of solid particles, the one describing contact electrification of metals is relatively mature. However, the process of contact electrification of other materials is rather complicated, not to mention the effects of surface state and surface contamination. Consequently, there is still lack of generally accepted theory for it at present.

5.1.4.1 Theory of contact electrification between metals

In daily life, electrostatic problems are mostly associated with insulating materials. Therefore, it is generally considered that electrostatic conducting materials, especially metal materials, will not produce static electricity. However, this is not the case. When contact-separation happens between two kinds of metals, transfer of electric charges also happens, thereupon static electricity is produced. The only thing is that during the separation, due to the good conductivity of metals, electrostatic charges are usually taken away quickly. After separation, there is little electrostatic charge remaining on metals, thus under normal conditions metals do not show any signal of carrying electrostatic charges.

1. Fermi Level and Work Function

There are large amounts of electrons inside metals. These electrons occupy different energy states that contain different energy in metals, which is called energy level distribution. According to theories of quantum mechanics, electrons always occupy lower energy levels in the initial state. At absolute zero, the highest energy level that electrons occupy in metals is called the metal Fermi level. In other words, energy states under metal Fermi level are all occupied at absolute zero while energy states above Fermi level are empty, namely no electrons on these energy states. Different metals have different Fermi levels.

Generally, metals are not at absolute zero. Besides potential energy, electrons of metals also possess kinetic energy. Electrons possessing larger kinetic energy are likely to jump to energy states above metal Fermi level. Thereupon, at room temperature, energy states above Fermi level are also occupied by a small amount of electrons while a small amount of energy states under Fermi level are free. In conclusion, metal Fermi level is the level that electron filling of metals can reach and the specific energy value that serves as the boundary of free energy states and full energy states. Since electrons participating in conduction or interacting with other substances are always possessing higher energy, it can be concluded that the energy (potential energy) possessed by free electrons of metals is the energy corresponding to metal Fermi level.

At room temperature, although free electrons of metals keep moving the thermal motion, they would not disengage from metals, indicating the existence of the force restricting electrons leaving the surfaces of metals. Such force is generated from two aspects. One is the effect of positive charges on metal lattice. The other one is that when free electrons approach the interface, they repel other electrons toward inner metal, generating excessive positive charges on metal surfaces to prevent electrons from leaving. From the energy point of view, the force is the forward electronic potential energy in metals. In other words, electrons of metals are in energy wells.

In order to escape from metals, electrons must possess enough energy. The minimum energy that can make electrons successfully escaping from metals is

Table 5.1 Work functions of some metal materials

Metal material	Work function
Silver	4.50 ~ 4.52
Copper	4.65
Aluminum	4.08
Iron	4.40
Gold	4.46
Nickel	5.03
Molybdenum	4.20
Tungsten	4.38

called the work function of metals. Unit of work function is electron volt, *i.e.* eV. Taking the potential energy of free electrons in vacuum as zero, the relationship between the work function of metals Φ and metal Fermi level u_F can be expressed as follow:

$$\Phi = -u_F \tag{5.1}$$

Work functions of metal materials can be obtained from experiments. At present, the more sophisticated experimental methods are thermionic emission method, luminescence method and standard metal method. However, values given by different methods are slightly different. Table 5.1 shows the work functions of some metal materials.

2. Contact Electrification of Metals

Two different kinds of metals, I and II, contact with each other. When the distance between is less than 2.5 nm, electrons inside them interchanges through the interface due to the tunnel effect of quantum mechanics. When balance is achieved, certain potential difference between the two metals is generated, as well as isometric different charges on both sides of the interface. If the two metals in contact are separated, they will respectively carry isometric different electrostatic charges, which is the process of contact electrification between metals.

It is assumed that Φ_1 represents the work function of metal I and Φ_2 represents the work function of metal II, and $\Phi_1 > \Phi_2$. Therefore, electrons of metal I are in the energy well with Φ_1 as its depth, and electrons of metal II are in Φ_2 deep energy well, as shown in Fig. 5.1. Since the potential energy of electrons in metal II is higher than that of I, when the two metal contacts with each other and the distance between them is less than 2.5 nm, electrons of the two metals will interchange through the interface. According to the lowest energy principle, metal II will have more electrons inflow to metal I until the Fermi level of them strikes balance. As a result, metal I gains electrons and is negatively charged on its surface, while metal II loses electrons and is positively charged on its surface.

Fig. 5.1 Work function of two metals before contact

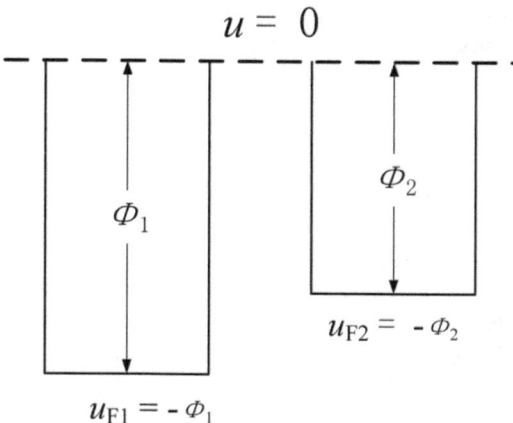

$$u = 0$$

$$u_{F1} = -\Phi_1$$

$$u_{F2} = -\Phi_2$$

It is supposed that the transfer of charges caused by contact between metals makes metal I and metal II respectively generates the electric potential U_1 and U_2 and $U_1 < 0$, $U_2 > 0$. Electronic potential of metal I increases by $-eU_1$, while electronic potential of metal II increases by $-eU_2$. Therefore, the potential of electrons in metal I can be expressed as:

$$W_1 = -\Phi_1 - eU_1 \tag{5.2}$$

The potential of electrons in metal II is:

$$W_2 = -\Phi_2 - eU_2 \tag{5.3}$$

When the exchange of electrons reach to balance, Fermi level of electrons in metal I and II equals the same, thermal energy of free electrons also equals the same. $W_1 = W_2$, as shown in Fig. 5.2.
Thus:

$$-\Phi_1 - eU_1 = -\Phi_2 - eU_2 \tag{5.4}$$

$$U = U_{21} = U_2 - U_1 = \frac{\Phi_1 - \Phi_2}{e} \tag{5.5}$$

Contact potential difference (hereinafter referred to as CPD) is also known as contact electromotive force. According to equations above, if $\Phi_2 < \Phi_1$, then $\Phi_2 - \Phi_1 < 0$, where e represents the absolute value of electronic quantity, thereupon $U > 0$, i.e. $U_2 > U_1$. Thus it can be indicated that when two metals contact with each other, the one with higher work function will be negatively charged while the one with lower work function will be positively charged. CPD of metals obtained from actual measurement is usually ranging from a few tenths of a volt to several volts.

Fig. 5.2 Potential of two
contact metals with
equilibrium state

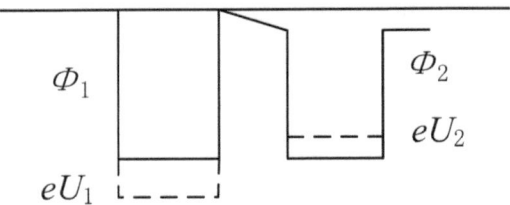

When two kinds of metal contact with each other, electric charges will only exist very close to the interface, forming the so-called electric double layer. It is assumed that d, representing the gap between electric double layers, is usually small. Therefore, the electric field strength between the gaps is considered uniform, which can be expressed as:

$$E = \frac{U}{d} \tag{5.6}$$

Thus the electric charge density of metal interface can be expressed as:

$$\sigma = \varepsilon_0 E = \varepsilon_0 \frac{U}{d} = \frac{\varepsilon_0}{ed}(\Phi_1 - \Phi_2) \tag{5.7}$$

Equation (5.7) has indicated that the electric charge density generated by the two metal is proportional to the difference between their work functions.

5.1.4.2 Contact electrification between metals and insulating materials

When insulating materials are contacting with metals, the transferring process of electrons between them not only depends on the Fermi levels of metals and energy states of electrons in insulating materials, but also the surface blemishes and surface contamination of insulating materials, even the time and frequency of the contact. Different from metals, insulating materials are generally unable to have definite Fermi levels and work functions. However, the equivalent work functions of insulating materials can be obtained through experiments. Table 5.2 shows the equivalent work function of some insulating materials. The polarity and quantity of electrification caused by contacts between metals and insulating materials can be decided according to the relative size of metal work functions and equivalent work functions of insulating materials.

5.1.4.3 Contact electrification of insulating materials

When two kinds of insulting materials contact with each other, transfer of electric charges will also happen. However, the mechanism of contact electrification of insulating materials is quite complicated. According to the types of electric carriers on which charge-transfer happens, the present theories can be divided into four categories: electron transfer model, ion transfer model, substance transfer model and aqueous solution ion transfer model. Contact electrification of insulating materials is obviously different to that of metals. Electric charges generated by

Table 5.2 Equivalent work function of some insulating materials

Material	Equivalent work function
Silver	4.50 ~ 4.52
Copper	4.65
Aluminum	4.08
Iron	4.40
Gold	4.46
Nickel	5.03
Molybdenum	4.20
Tungsten	4.38

metal contact electrification are only distributed on metal surfaces. However, when it comes to the contact electrification of insulating materials, electric charges are not only distributed on surfaces, but also inside the materials, forming a volume distribution of charges.

All materials, especially crystal-structured substances, are different in structures of their crystal faces and interiors. On the surfaces, the periodic array of microscopic particles, or the long range order and potential periodicity, etc., is truncated, which forms the "dangling bonds" and results in surface energy levels. Appearance of these surface energy levels will seriously affect characteristics of contact electrification of insulating materials. Due to the surface energy levels, gaining and losing of electrons will happen on the surface, which equals to "inserting" a thin metal plate into the surfaces in contact with each other. Therefore, it will directly affect the contact electrification theory of solid mentioned before. Besides, generally, object surfaces are contaminated with other substances which will affect the result of contact electrification.

Electrons of electrified insulators are generally considered electrified charge carriers. Except for substances with simple crystal structures such as ionic crystal, the cognition of energy levels and work functions of insulating materials is deficient both theoretically and experimentally. Electrified charge carriers involved in surface conduction of insulators, especially when the surface adsorbs water, can be considered to be ions. When insulators are electrified, some scholars would consider the electrified charge carriers moving through the contact interface as ions. Even if the air is dry, there is still water adsorbed on the surface of objects. Adsorption capacity varies as object differs. In low humidity conditions, water is adsorbed with monolayer; in high humidity conditions, it is adsorbed with multilayer. Sometimes objects can absorb water molecules, and a part of these water molecules dissociated into H+ and OH−.

5.1.4.4   Separation process

From discussion above, it can be inferred that when two objects contact with each other, charge exchange happens on the contact interface. When it reaches an equilibrium state, electric double layer is generated on the interface, and the two objects carry equal positive and negative charges respectively. Since the spacing of electric double layer is very small, if the two objects in contact with each other are considered as one system, the system is still uncharged and demonstrates electric neutrality. Only after contact and separation can result in the two objects carry positive and negative charges respectively. Moreover, after separation, the absolute value of charges carried by each object does not equal to that of positive or negative charges in electric double layer when the objects are contacting and in balance. Generally, the former one is smaller than the latter one. If Q represents the absolute value of charges carried by any object after separation, and Q_0 represents the absolute value of positive or negative charges in electric double layer before separation, their relationship can be expressed as:

$$Q = kQ_0 \qquad\qquad (5.8)$$

In which the range of k is $0 < k < 1$, i.e. $Q < Q_0$. The reason why Q is smaller than Q_0 is that during the separation, some of the charges in electric double layer display the phenomenon of dispersion. Value of k is directly related to the physical phenomena such as separation velocity of the two objects in contact with each other, surrounding environment of contact interface, field-induced emission during the separation and gas discharge.

1. Charge Dispersion of Contact Interface

The charge dispersion of contact interface happens in early separation. When the separation begins, the tunnel effect is still effective, which makes electrons still capable to move. Therefore, even during the separation process, electrons can move within the effective interval of tunnel effect. When the distance between two objects D is less than the critical distance of tunnel effect d_C, the potential difference of two metals will remain as the original value. As capacitance decreases, surface charge density also decreases. This is called the countercurrent of electric charges. If the interval exceeds the critical distance of tunnel effect d_C, potential difference between objects will no longer remain constant. If charges can no longer move through the contact interface except for dispersion, then electrified objects will hold constant quantity of electricity. As capacitance decreases, electric potential U increases.

2. Field-Induced Emission

When very strong electric fields are acting on surfaces, even it is at relatively low temperature; objects are still capable to emit electrons. Such phenomenon is called field-induced emission. Actually, there are bumps more or less on contact interfaces. During the separation, electric charges are gathered together at these bumps, probably forming strong electric fields and leading to field-induced emission. Electrons emitted in field-induced emission will neutralize the charges between contact interfaces during separation, which decreases the value of k in Eq. (5.8).

3. Charge Leakage Through the Surface and Interior of the Object

No matter conductors or insulating materials, objects can conduct electric charges to some extent. During the process of separation electrification, conduction effect, through which charges are leaked into the ground from the surface and interior of the object, makes the charge density of electric double layer decrease. Capacity of electric leakage mainly depends on the quantity of bleeder resistance of object surface and interior simultaneously, and it also depends on the leakage through contact interface during the early separation process of objects. Since the contact interface is not an ideal plane, in early separation, electric potential of the separation point increases, while electric charges gather together to the places with lower potential and points that are not separated yet, namely gather together to the bumps. Thus, it makes separated charges leak through the points that are not separated yet.

Above analysis indicates that value of k in Eq. (5.8) depends on the value of electrostatic bleeder resistance of objects. The larger the electrostatic bleeder resistance is, the faster the separation is, and the closer the value of k is to 1; conversely, the closer the value of k is to 0. This is one of the main reasons that objects carry lesser charges in contact electrification of metals while they carry more charges in contact electrification between high polymer materials with good insulating properties and other high polymer materials.

4. Influence of Gas Discharge

When the leakage resistance is large, charges of insulator cannot disperse through the surface and interior of objects. Discharge of surrounding gas is the main factor that determines the maximum value of electricity. During the separation, the electrostatic field strength between electric double layers may exceed the breakdown strength of space air, which will lead to gas discharge. Gas discharge reduces the quantity of charges carried by electrified bodies themselves. When it happens between the interfaces, the space distance can be estimated by Paschen's law. According to Paschen's law, there exists a minimum value of voltage that can make gas discharge happen. This minimum value depends on the product of air pressure and the space distance of air. When the air pressure is 1 mmHg, the corresponding discharge space is 6 μm; while when the air pressure is 1 atm, the corresponding discharge space is 8 μm.

5. The Role of Friction in Contact Electrification

At the very beginning, people referred to the electrostatic electrification as triboelectrification. In fact, friction is not the necessary condition to trigger electrostatic electrification. A simple process of contact-separation would electrify objects; however, friction does increase the effect of contact electrification. The process of friction is actually the continuous process of contact and separation between different contact points on contact interface of two objects rubbing together. For metal conductors, only the final moment of separation between two conductors contributes to electrostatic electrification; for insulators, the whole friction process is related to electrostatic electrification. For instance, the elevation of temperature caused by friction, the fracture of the bumps on material interfaces, thermolysis, piezoelectric effect and pyroelectric effect and so on will change the quantity of electrostatic electrification. Besides, factors including friction type (such as symmetric friction or asymmetric friction, friction along fixed direction on plane or rotary/twisting friction), friction time (instantaneous contact separation or separation after long-period friction), friction velocity (separation velocity), contact area of friction (length of friction) and positive pressure of friction are all related to the electric quantity of electrification.

5.2 Basic Theory of Particle Electrification

5.2.1 Estimate the Charges of Powder Electrification

The main mechanism of the powder electrification is caused by the friction, collision, and separation between the powder and the pipeline, the wall, as well as the belt, and the mutual friction, collision and separation, solid fracture, fragmentation among the powder particles themselves and other factors under the motion state, such as fast flow, jitter or vibration. These phenomena may appear at any time in the powder transport, spray, mixing, crushing, screening and other processes.

1. The powder electrification parameters of gas transport

   Taking advantage of the powder electrification parameters in the air transport (pneumatic conveying), including charge current, volume charge density, mass charge density and so on. The following formula is given in the relevant data:

① *Charging current (electrification current) density*

   When powder pneumatic conveys, the powder and the pipeline have the same amount of charge respectively, and generate charge current. Charge current density is calculated as:

$$J = \frac{0.35 K_C^{0.4} v^{1.8} K_m^{1.8} K_s K_t \sigma_m}{d \left(\lg \frac{Re}{8} \right)^{1.8}} \tag{5.9}$$

where,

d diameter or equivalent diameter of powder particle, m;

v the flow rate of gas, m·s^{-1};

Re   Reynolds number;

K_C coefficient determined by powder material and pipe material, s^2·m^{-2};

K_m ratio of the maximum of pulsating velocity to the maximum of the fluctuate velocity of airflow, i.e. $K_m = \sqrt{0.75\alpha\frac{\rho_g}{\rho_f} \cdot \frac{D}{d}}$

Where,

α coefficient of pulse damping, for spherical particles, α take 0.44; for non-spherical particles, $\alpha = \frac{50.66\eta^{\lg\frac{Re}{8}}}{dv} + 0.48$;

ρ_g the density of the gas used to transport the powder, kg·m^{-3};

ρ_f the density of powder material, kg·m^{-3};

K_s differential coefficient, which is the ratio of the longitudinal velocity of the airflow to the longitudinal velocity of the powder;

K_t volume concentration of powder, $K_t = \frac{V_t}{V_t+V_g}$, where V_t is the flow rate of powder, which is the volume of powder transported per second; where V_g is the gas flow, that is, the volume of gas delivered per second;

σ_m charge density.

② *The saturated volume charge density of powders ρ_∞*

The empirical formula for the saturated volume charge density in pneumatic conveyance is given

$$\rho_\infty = 19.5\beta^{-0.74}v^{1.18} \tag{5.10}$$

where,

v gas velocity, m·s^{-1};

β gas-solid ratio, that is, the powder weight suspended in per cubic meter of air, kg·m^{-3}.

③ *The mass charge density and saturation mass charge density of powder*

The mass charge density of powder indicates the charge quantity of the powder, which is an important parameter to characterize the degree of powder charged. With other conditions unchanged, the mass charge density of powder will increase with the increase of charging time. Figure 5.3 gives an example of mass charge density versus charge-time curves for different gas-solid ratios.

The mass charge density is called the saturation mass charge density when $t \to \infty$. The saturated mass charge density of the powder is determined by the charge density of the saturated volume and gas-solid ratio, and according to the following formula:

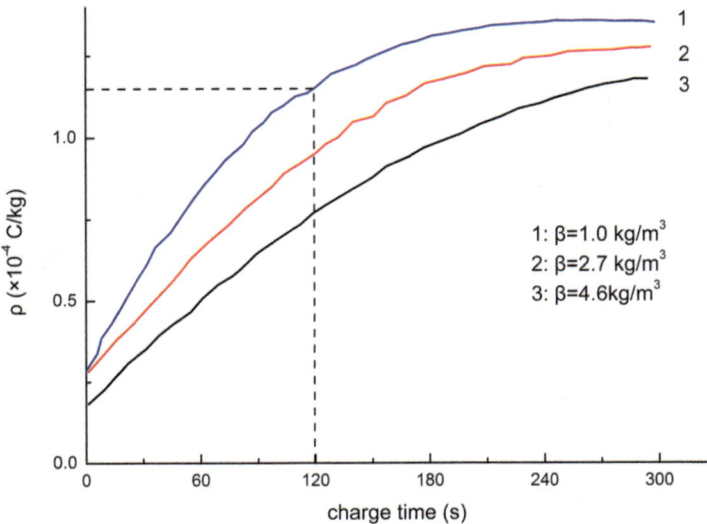

Fig. 5.3 Relationship of mass charge density and charge-time curves for different gas-solid ratios

$$\rho_{m\infty} = \rho_\infty \beta^{-1} \qquad (5.11)$$

where,

ρ_∞ saturated volume charge density of powder, $C \cdot m^{-3}$;

β gas-solid ratio, $kg \cdot m^{-3}$;

$\rho_{m\infty}$ saturated mass charge density of powder, $C \cdot kg^{-1}$;

The mass charge density of the powder is calculated as follows:

$$\rho_m = \rho_{m\infty}\left(1 - ke^{-\frac{t}{\tau}}\right) \qquad (5.12)$$

where,

$\rho_{m\infty}$ saturated mass charge density of powder, $C \cdot kg^{-1}$ (seen in the formula Eq. (5.12));

k coefficient determined by material and shape of powder and pipe, $0 < k < 1$;

τ time constant of powder media, $s; \tau = \varepsilon\rho$, where, ε is the permittivity of the powder medium, F/m, ρ is the volume resistivity of the powder, $\Omega \cdot m$.

④ *The initial accumulation rate I_{mo} of powder charge*

The empirical formula is:

$$I_{mo} = -2.47v^{0.4}\beta^{-1.07} \qquad (5.13)$$

Where, v is the initial rate of charge accumulation, $\mu C \cdot kg^{-1} \cdot s^{-1}$.

2. The limit mass charge density of suspended powder

The powder material in the suspended state in the gas cannot be charged beyond a certain limit. Otherwise, the field strength of each powder particle surface is strong enough to ionize the surrounding atmosphere, resulting in electrostatic discharge. The limit mass charge density of suspended powder is calculated as:

$$\rho_{mm} = \frac{3\sigma}{\rho_M r} \tag{5.14}$$

where,

ρ_{mm} limit mass charge density of powder, $C \cdot kg^{-1}$;
σ surface charge density;
ρ_M the density of the powder particles, $kg \cdot m^{-3}$;
r the radius of powder particle, m;

5.2.2 Charge Transfer Generated by the External Electric Field

When the symmetry between the contacting surfaces would appear to preclude a driving force for charge transfer, how is charge transfer possible for particles with identical chemical composition? Pähtz and co-workers (Pahtz et al. 2010) point out that external electric fields break this symmetry, thus providing a driving force for charge transfer. Their basic idea is very simple. In an external dipolar electric field, which for argument's sake is taken to be oriented with the negative pole up, a particle becomes polarized such that its top is locally negative and its bottom is locally positive. When two such particles collide in a top-bottom orientation, the locally positive surface of the higher particle contacts the locally negative surface of the lower particle, creating a driving force for charge transfer such that the higher particle gains net negative charge and the lower particle gains net positive charge, as shown in Fig. 5.4. As the external electric field asserts a spatial distribution of particles where net positive particles are on average higher than net negative particles, subsequent collisions make the positive particles become even more positive, and the negative particles even more negative.

5.2.3 Charge Transfer Generated by Asymmetry Contact

Lowell (Lowell and Truscott 1986a, b) noted that triboelectrification of identical insulator was produced by the asymmetry contact between two contacting surfaces, and he inferred the number of high-energy electrons transferred from one contacting surface to the other one was different in the two contacting surfaces.

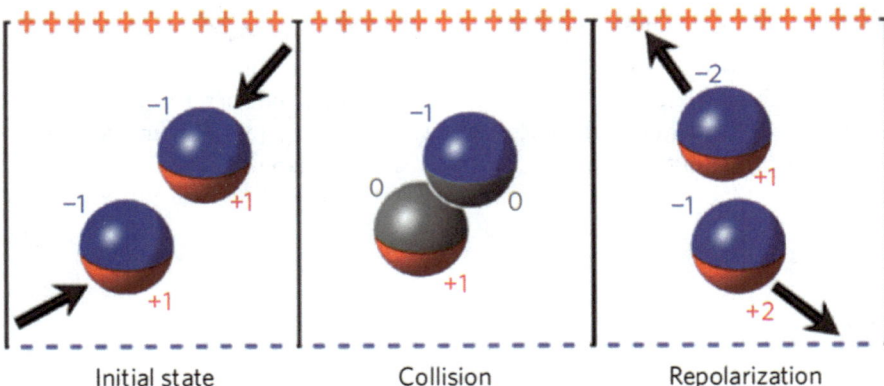

Fig. 5.4 The proposed charging mechanism of colliding particles in an electric field. Initially (*left panel*) a pair of particles polarized by an external electric field collide (*central panel*) to neutralize adjoining hemispheres. Once separated (*right panel*), the particles again become polarized by the external field. In this way, initially neutral but currently polarized particles gain one unit of charge following every collision. *Blue* denotes negative and *red* positive charge, as indicated by the numbers beside each hemisphere, and the arrows indicate representative particle velocities (Pahtz et al. 2010)

A simple model to address the triboelectric charging in identical material particles systems is proposed by Lacks and Levandovsky (2007). The basis of the model is the existence of electrons trapped in high-energy states, which can be released during collisions with another particle and transferred to the other particle. This model shows that the triboelectric charging in insulator systems composed of particles of identical material can be attributed to a distribution of particle sizes, such that smaller particles tend to charge negatively and larger particles tend to charge positively.

The physical basis of the theory is that electron states in insulators are spatially localized, so that electrons can be trapped in high-energy states even if vacant lower energy states exist elsewhere in the material, as shown schematically in Fig. 5.5.

A binary mixture of small (S) and large (L) particles is considered. There are N particles, and the number fractions of small and large particles are x_S and x_L, respectively ($x_S + x_L = 1$). The radii of the particles are R_S and R_L ($R_S < R_L$).

All particles initially are assumed to have identical surface density of trapped high energy electrons, ρ_0 electrons/(unit surface area). Thus there are initially $N x_S 4\pi R_s^2 \rho_0$ trapped high energy electrons on the small particles, and $N x_L 4\pi R_L^2 \rho_0$ trapped high energy electrons on the large particles.

These trapped high energy electrons are released by collisions, and transferred to a low energy state in the colliding particle. Each particle collides with both large and small particles, and thus distributes its trapped high energy electrons to low energy states in both large and small particles. The fraction of the collisions undergone by a small particle involving a large particle is represented by $f_{S,L}$, and

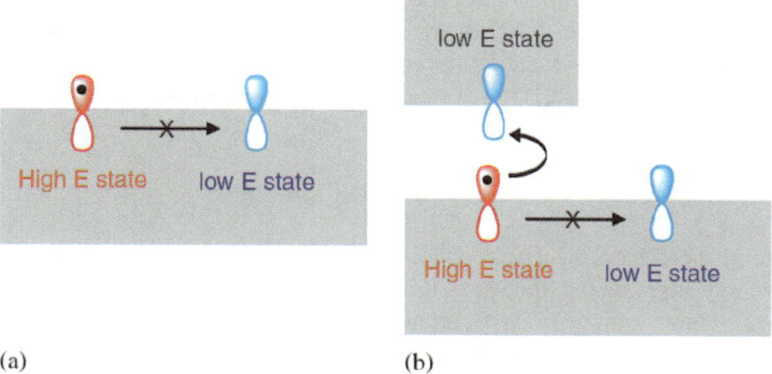

(a)   (b)

Fig. 5.5 Schematic representation of the process of electron transfer upon contact of two insulators (Lacks and Levandovsky 2007). (**a**) Electrons on the surface of an insulator can be trapped in high-energy states. (**b**) These electrons can be released from the high-energy states when a collision brings a low-energy state on another particle in close proximity

the fraction of the collisions undergone by a large particle that involve a small particle is represented by $f_{L,S}$.

The particles are initially neutral, therefore the charge on a particle corresponds to the number of electrons lost minus the number of electrons gained. The average charge on the small particles, Q_S, is then given by $Q_s = \frac{N_{XS}4\pi R_S^2 \rho_0 f_{S,L} - N_{XL}4\pi R_S^2 \rho_0 f_{L,S}}{N_{XS}}$

$$Q_s = -4\pi R_S^2 \rho_0 \left(\frac{x_L}{x_S}\right)\left(1 + \frac{R_L}{R_S}\right)$$

$$\times \left\{ \frac{\left(\frac{x_L}{x_S}\right)\left(\frac{R_L}{R_S}\right)^2 \left[\left(\frac{R_L}{R_S}\right)^2 + 2\left(\frac{R_L}{R_S}\right) - 3\right] + \left[3\left(\frac{R_L}{R_S}\right)^2 - 2\left(\frac{R_L}{R_S}\right) - 1\right]}{\left[4 + \left(\frac{x_L}{x_S}\right)\left(1 + \frac{R_L}{R_S}\right)^2\right]\left[4\left(\frac{x_L}{x_S}\right)\left(\frac{R_L}{R_S}\right)^2 + \left(\frac{x_L}{x_S}\right)\left(1 + \frac{R_L}{R_S}\right)^2\right]} \right\}$$

$$(5.15)$$

Thus, this analysis shows that $Q_S \leq 0$. The equality holds when all particles are of the same size and the small particles are charged negatively when the particles are different sizes.

For granular systems of chemically identical insulators, Lacks et al. (2008) showed that a systematic charge transfer is produced by a different asymmetry. They also showed that, after several initial collisions in which small and large colliding particles lose roughly equal amounts of trapped electrons, smaller particles have nonetheless lost a larger fraction of their trapped electrons than larger particles. Therefore, in subsequent collisions, smaller particles give up fewer trapped electrons than larger particles do, making the smaller particles to be charged negatively and larger particles to be charged positively.

For simplicity, Kok and Lacks (2009) assume that electrons on the surface of these particles can be in either a low-energy (L) or a high-energy (H) state. The number of high- and low-energy electrons on the particle's surface at time t are denoted as n_{iH} (t) and n_{iL} (t), and the initial number of high-energy electrons is given by

$$n_{iH}(0) = 4\pi R_i^2 \rho_{H,0} \tag{5.16}$$

During a collision, electrons can relax from high-energy states on one particle to low-energy states on the opposite particle is assumed. Specifically, all high-energy electrons within a distance δ_0 of the surface of the opposite particle (Fig. 5.6) will tunnel to empty low-energy states on that particle's surface. Lowell (1979) modeled this electron transfer process in terms of the tunneling dynamics of a particle in a one-dimensional square well separated by an energy barrier from another square well. It is showed that the maximum distance δ_0 that an electron in the ground state of a square well can tunnel during a collision is approximately given by

$$\delta_0 = \frac{\hbar}{\sqrt{8mE_b}} \ln\left(\frac{\hbar\, t_{coll}}{\eta m\, a^2}\right) \tag{5.17}$$

Where $\eta = (1.12/\pi^2)(2 + \pi)$, E_b is the height of the energy barrier between the two potential wells, m is the electron mass, $\hbar$ is the reduced Planck constant, t_{coll} is the time scale of the collision, and a is the radius of the well corresponding to the electron trap. Since electrons can transfer between particles during collisions, net charges can develop on the particles. The net charge on a particle of type i is then given by

$$q_i(t) = e[n_{iH}(0) - n_{iH}(t) + n_{iL}(0) - n_{iL}(t)] \tag{5.18}$$

It is hypothesized that this transfer of electrons proportional to the particle's surface area, at the instant of collision, is within the distance δ_0 of the surface of the opposite particle. The number of electrons transferred from the high-energy states of particle i ($-\Delta n_{iH}$) to the empty low-energy states of particle j (Δn_{jL}) is equal to

$$-\Delta n_{iH} = \Delta n_{jL} = 2\pi \rho_{H,0} R_i^2 (1 - \cos\theta_i) \tag{5.19}$$

The angle θ_i represents the maximum angle from the contact point for which the surface of particle i is within the distance δ_0 of the surface of particle j and satisfies

$$\begin{aligned} R_i \sin\theta_i &= \left(R_j + \delta_0\right)\sin\alpha_j \\ R_i \cos\theta_i + \left(R_j + \delta_0\right)\cos\alpha_j &= R_i + R_j \end{aligned} \tag{5.20}$$

where, the angle α_j is defined in Fig. 5.6. Solving Eq. (5.20) for α_j, using that $\cos(\arcsin x) = \sqrt{1 - x^2}$, squaring both sides, and solving for $\cos\theta_i$, we obtain

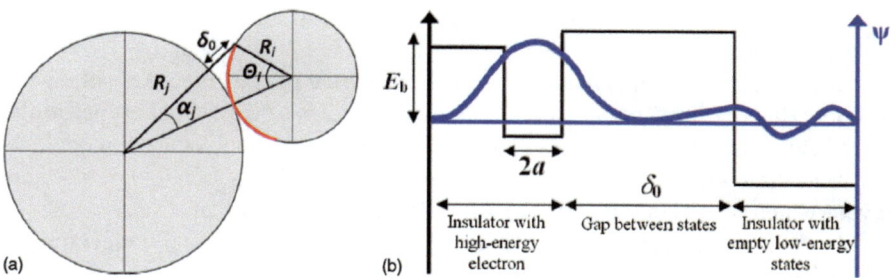

Fig. 5.6 (**a**) Schematic of the charge transfer occurring during a collision between two spherical particles of identical material but of different sizes R_i and R_j. The angle θ_i represents the maximum angle from the point of contact on particles i and j from which trapped high-energy electrons can transfer to empty low-energy states on the opposite particle. The area in which electrons can transfer in this manner is indicated by the thick *red* arc. (**b**) Simplified schematic representation of the wave function of a high-energy electron in an electron trap near the surface of another insulator with empty low-energy states. The thin *black* line denotes the electron's potential energy as a function of position and the thick *blue* line denotes its wave function

$$\cos \theta_i = 1 - \frac{\delta_0}{2R_i}\frac{2R_j + \delta_0}{R_i + R_j} \tag{5.21}$$

Substituting this result then yields the number of electrons transferred between the colliding particles in terms of the particle sizes, the density of trapped states, and the tunneling distance

$$-\Delta n_{iH} = \Delta_{jL} = \pi \rho_{H,0} R_i \delta_0 \frac{2R_j + \delta_0}{R_i + R_j} \tag{5.22}$$

By substituting Eq. (5.21) into Eq. (5.18). The net charge transfer experienced by particle i is obtained as

$$\Delta q_i = \pi e \rho_{H,0} \delta_0^2 \left(\frac{R_i - R_j}{R_i + R_j} \right) \tag{5.23}$$

Simple geometry thus makes the larger colliding particle obtain a positive charge denoted by Eq. (5.23), while the smaller particle loses a charge of the same magnitude.

However, when the ratio of two particles radii R_i/R_j is in the certain range, the net charge transfer will be less than one elementary charge e for a single collision event when the two contact particles have the identical initial densities of high-energy trapped surface states. Kok and Lacks (2009) adopted the hard sphere model in their charging mechanism to deal with the collision process between two particles, and concluded that the net charge transfer is independent of the relative collision velocity. However, the hard sphere model is an instantaneous point contact considering pre- and post-collision velocities of colliding particles without the contact details of collision bodies.

In order to get the details of the collision, i.e., the contact area, the soft sphere model (Hu et al. 2012) which considers sand particles as elastic bodies deforming each other during a collision is used. Therefore, in the soft sphere model, the asymmetric contact area is related to the collision velocity of a single collision event which results in the net charge transfer not being constant.

By denoting, respectively $S_{n,i}$, S_τ and $S_{\omega,i}$ as the arcs of normal, tangential and circumferential directions obtained from the soft sphere model, the maximum tunneling arc $S_{i,max}$ in soft sphere model can be written as $S_{i,max} = \max(S_{n,i}, S_\tau) + S_{\omega,i} + S_{0,i}$, in which $\max(S_{n,i}, S_\tau)$ is the maximum arc between the normal and tangential directions which is chosen to avoid the overlap part of the contact arc length; $S_{0,i}$ is the tunneling arc of particle i in the hard sphere model. It can be found that the tunneling distance $S_{i,max}$ in the soft sphere model is B_i times bigger than that in the hard sphere model, where the coefficient $Bi = S_{i,max}/S_{0,i}$. Therefore, the numbers of electrons trapped in high-energy states tunneling from particle i to particle j and from particle j to particle i are

$$\begin{aligned} N_H^i &= B_i \frac{\pi e \delta_0 R_i (2R_j + \delta_0)}{R_i + R_j} \rho_{iH,0} \\ N_H^j &= B_j \frac{\pi e \delta_0 R_j (2R_i + \delta_0)}{R_i + R_j} \rho_{jH,0} \end{aligned} \tag{5.24}$$

where $\rho_{i/jH,0}$ is the initial density of high energy trapped surface states. The coefficient $B_{i/j}$ is related to the relative impacting velocity, the relative impacting angle and the particle size in the soft sphere model. The net charge transfer of particle i is written as

$$\Delta q_i = N_H^i - N_H^j \tag{5.25}$$

Note that when the coefficients $B_i = B_j = 1$, the net charge transfer, Δq_i, is equivalent to that given by the triboelectric charging model proposed by Kok and Lacks (2009) for a single collision between two particles which have identical initial densities of high-energy trapped surface states $\rho_{H,0}$.

Desch and Cuzzi (2000) proposed a model in which the collision charge transfer depends on the preexisting charges, the particle sizes, and the difference in the particles' contact potential. They proposed that

$$\begin{aligned} q_S' &= C_1(q_S + q_L) - C_2 \Delta\Phi, \\ q_L' &= (1 - C_1)(q_S + q_L) + C_2 \Delta\Phi, \end{aligned} \tag{5.26}$$

where q_S and q_L are the charges of the smaller and larger particles before the collision, q_S' and q_L' are the charges after the collision, $\Delta\Phi$ is the difference in particle contact potential, and C_1 and C_2 are functions of the mutual capacitances (and thus the radii) of the two particles. For particles of similar composition (i.e., $\Delta\Phi = 0$), such as typical soil particles, Eq. (5.26) suggests that no charge transfer occurs when the colliding particles are not initially charged, which contradicts observations. To mitigate this problem, an effective contact potential difference between particle pairs of similar composition but different sizes is proposed, that is (Kok and Renno 2008)

$$\Delta\Phi_{eff} = S(r_L - r_s)/(r_L + r_S) \tag{5.27}$$

where S (in volts) is a physical parameter that scales the collisional charge transfer, and r_S and r_L are the radii of the small and large particles, respectively. This simple model has a functional form consistent with observations - smaller particles acquire net negative charge during collisions with larger particles, and the charge transfer is reduced as the relative difference in particle size decreases. $S = 6 \pm 4$ V

Based on the research reported by Grzybowski's group, a statistical model to compute the net charge of electrified particle after a single collision and multiple collisions are established, respectively (Xie et al. 2013).

After collision with the incident particle for the first time, the net charge of the target particle is calculated by

$$\Delta Q_2(1) = \alpha(1)\rho_1 N_{D1}(0)\frac{N_{A_2}(0)}{N_2} - \alpha(1)\rho_2 N_{D2}(0)\frac{N_{A_1}(0)}{N_1} \tag{5.28}$$

The original spherical caps of the incident sphere and target sphere by S_1 and S_2, whose areas are A_1 and A_2, respectively. It is assumed that the numbers of surface charge donors on S_1 and S_2 are $N_{D1}(0)$ and $N_{D2}(0)$, respectively, and the numbers of acceptors are $N_{A1}(0)$ and $N_{A2}(0)$, respectively, before collision.

According to the theory presented by Grzybowski's group, $\alpha(1)$ denotes the probability that charges can be transferred from a donor on S_1 to an acceptor on S_2, or in the reverse direction. For the value of $\alpha(i)$, Grzybowski et al. (2005) gave the following formula, $\alpha(i) = 1 - \sum Q(i-1)/Q_s$, where $Q(i-1)$ is the net surface charge before the ith collision and Q_s is the saturated net surface charge. Thus, for $i = 1$, $\alpha(1) = 1$. ρ_1 is the amount of charge moving from a donor on S_1 to an acceptor on S_2, while ρ_2 denotes the amount of charge moving from S_2 to S_1 when the two donor surfaces contact during a single collision process. Let N_1 and N_2 be the total numbers of charge donors and acceptors on S_1 and S_2 in initial state, i.e., without any contact between the two particles, respectively, which will get involved in contact during a collision process. Obviously, $N_1 = N_{D1}(0) + N_{A1}(0)$ and $N_2 = N_{D2}(0) + N_{A2}(0)$. Here, the area of a single donor is assumed to be equal to that of an acceptor. Thus, by letting A_0 be the area of a single donor or acceptor and ρ be the charge density of donors and acceptors, $N_1 = A_1/A_0$, $N_2 = A_2/A_0$, and $\rho_1 = \rho_2 = \rho A_0$. Consequently, we obtain the following formula to compute the net charge of the target particle:

$$\Delta Q_2(1) = \alpha(1)\rho A_0 \left[N_{D2}(0) - N_{D1}(0) + A_0 N_{D1}(0) N_{D2}(0) \left(\frac{1}{A_1} - \frac{1}{A_2} \right) \right] \quad (5.29)$$

Both the number of donors and acceptors are stochastic variables and obey the binomial distribution. Therefore, the net charge of the target particle after the first collision is also a random variable, of which the mean value can be obtained by taking average on both sides of Eq. (5.29)

$$\langle \Delta Q_2(1) \rangle = \alpha(1)\rho A_0 \left[\langle N_{D2}(0) \rangle - \langle N_{D1}(0) \rangle + A_0 \langle N_{D1}(0) N_{D2}(0) \rangle \left(\frac{1}{A_1} - \frac{1}{A_2} \right) \right]$$
$$(5.30)$$

Where $\langle \cdot \rangle$ denotes the first order moment of a random variable.

Assuming the probability for a position on the surface to be a donor be p_D, the probability for it to be an acceptor will equal to 1- p_D. Here, it is assumed that the probability, p_D is the same for all the positions on the surface. Therefore, $\langle N_{D_1}(0) \rangle = p_D N_1 = p_D A_1 / A_0$ and $\langle N_{D_2}(0) \rangle = p_D N_2 = p_D A_2 / A_0$. There is no relationship between the numbers of donors on the surfaces of the two particles, therefore, $N_{D1}(0)$ and $N_{D2}(0)$ are independent from each other. So, $\langle N_{D_1}(0) N_{D_2}(0) \rangle = \langle N_{D_1}(0) \rangle \langle N_{D_2}(0) \rangle = (p_D)^2 A_1 A_2 / (A_0)^2$. By inserting this formula into Eq. (5.30), the net charge of the target particle after the first collision with the incident particle can be determined by

$$\langle \Delta Q_2(1) \rangle = \alpha(1)\rho p_D (1 - p_D)(A_2 - A_1) \quad (5.31)$$

$\langle \Delta Q_2(1) \rangle$ is named the initial net charge of the target particle, in the sense that it is produced by the first collision between the two particles.

The polarity of the net charge carried by either of the particle is dependent on the polarity of charge carriers transferred between the two particles and the relative size, or more precisely, the sign of $(A_2 - A_1)$. That means the larger particle can be positively charged or negatively charged, depending on their materials and sizes. Obviously, when collision occurs between two particles of the same size, neither of them will carry net charge.

5.2.4 Charge Transfer from Aqueous Ion Shift on Particle Surfaces

Liquid can wet the surface of the solid. The main reason is that the solid surface can attract the liquid molecules. Macroscopically, it is determined by the characteristics of liquid molecules and the free energy level of the solid surface. From the microscopic aspect, sand adsorption of water is mainly determined by the interaction force between the sand particles surface and water molecules. For the first layer of water molecules adsorbed, short-range force plays a major role, while for the two or more layers of water molecules, long-range interaction plays a major role. In addition, the surface structure of sand particles also has an important effect on the ability of sand to adsorb water. The more uneven the surface structure of sand and the higher the polarity is, the higher the surface energy and the greater the adsorption capacity of water is. Therefore, there are many factors influencing sand adsorption capacity of water, including the surface properties of sand, temperature, the concentration of the surfactant contained in water, and the distance between water molecules and the sand surface.

In order to verify the ability of sand particle surface to adsorb water molecules, the water absorption of sand particles was tested.

1. The microscopic morphology of sand particles

The tested sand particles were sampled in the Kumtag desert. The sands were thoroughly washed with an acetone solution in an ultrasonic cleaner and then dried. The dried sample was placed on a transparent adhesive tape. Then the surface of sand was coated with a layer of gold powder by using the standard vacuum coating machine to facilitate the conductivity, and sprayed gold samples were attached to the sample pier, and placed in the scanning electron microscope for observation. Figure 5.7 is a typical micro-morphology of the sand surface. In addition to the rough surface and the pothole Kankan, there are many different sizes of pores and uneven holes on the sand surface. In fact, the ability to adsorb water molecules is related to the surface of the pores, and the water molecules first occupy the pores with small size, and then are gradually adsorbed on the surface of all the holes. Thus, the ability of sand to adsorb water is related to the surface area of pores present on its surface, as shown in Fig. 5.7c. There are pores of various sizes ranging from nanometers to micrometers. These holes make sands adsorb water molecules in the atmosphere, and form water molecules layer on the surface.

2. Water absorption test of sand particles

In order to test the water absorption performance of the sand samples, the sand was placed in the oven at 105 °C baking temperature for 6 h to completely dry the absorbed water contained in them. Then sand particles were divided into two groups, ten samples, each containing 100 g of sand, and they were placed in room temperature and humidity control box respectively.

The temperature and humidity curves in the room temperature environment are shown in Fig. 5.8. At room temperature, the range of temperature change is large, but the fluctuation is small, and it basically has a smooth rise and fall tendency. The humidity at room temperature also maintained a stable trend, but the variation range was small. It can be seen in Fig. 5.8 that the change of relative humidity is closely related to temperature. The relative humidity decreases with the increase of temperature. When the temperature is at the maximum, the relative humidity is the smallest. After that, the temperature decreases and the relative humidity do not change much.

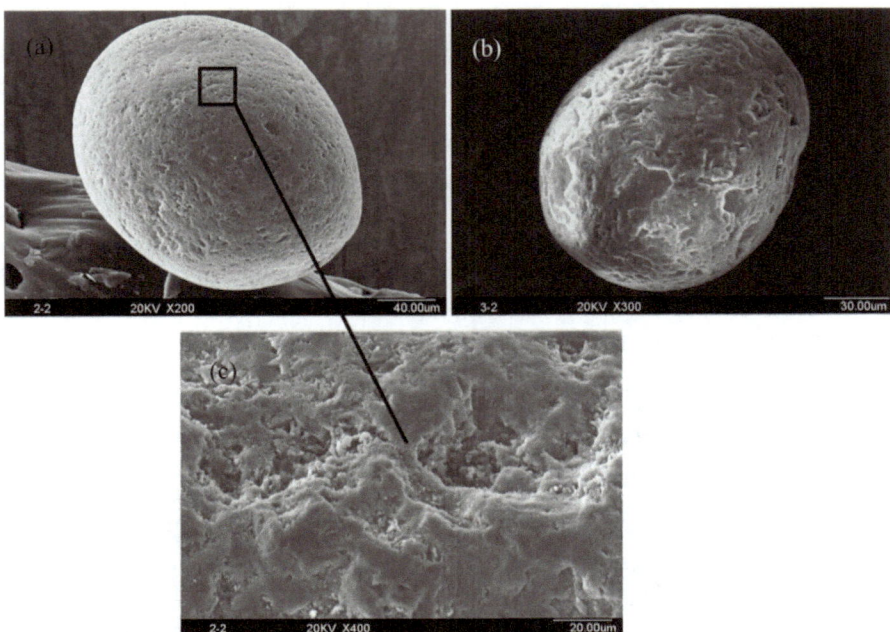

Fig. 5.7 Micro-morphology of sand particles. (**a, b**) The SEM image of sand surface, ×200. The sand surface is microporous, there are many irregular zones. (**c**) An enlarged image of (**a**) showing details on the sand surface, ×400. There are many nanoscale pores on the pit zones, which could adsorb water from atmospheric environment

Fig. 5.8 Daily change curve of temperature and humidity in the room temperature environment

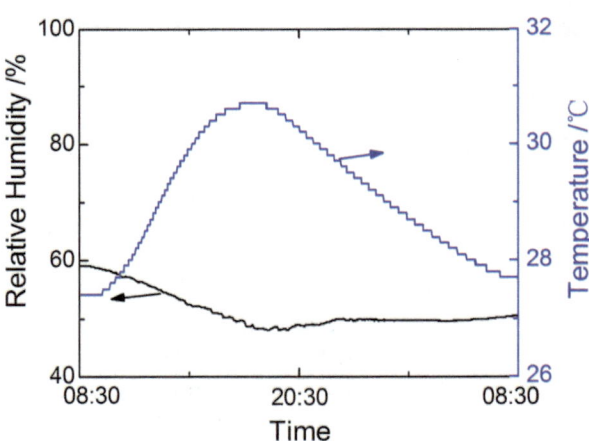

Control the temperature inside the box at 30 °C by using the control system of the incubator, and then use the humidifier to humidify the box, control the humidity in the box by adjusting the amount of humidifier, Fig. 5.9 shows the change curve of the temperature and humidity inside the box. As the incubator uses air-conditioning system to control the temperature inside the box, when the temperature reaches the set target, the air conditioning system will automatically shut down; when it is below a certain value, the air-conditioning system will automatically open, so the

Fig. 5.9 The daily change curves of temperature and humidity. Control target: temperature: 30 °C; relative humidity: 80%

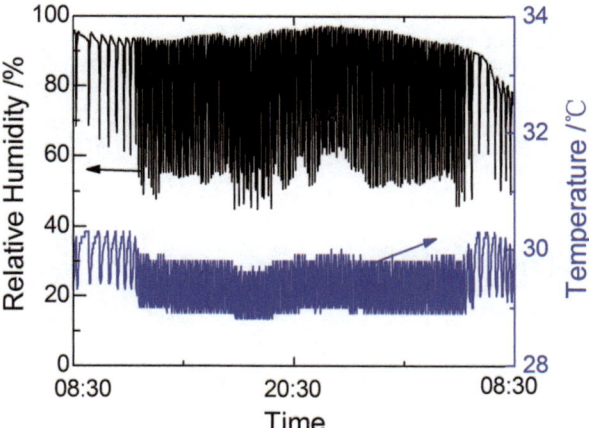

temperature fluctuations within the box is large, resulting in severe fluctuations of humidity inside the box.

The two groups of samples were placed in the above two environments, and the weight changes of sand were recorded at a regular interval. It can be considered that the weight change of sand particles is due to its adsorption of water in the environment. Taking the average weight change of each group of ten samples, we can get the water absorption curve of sand under different temperature and humidity, as is shown in Fig. 5.10. The temperature and humidity in the figure takes the average value in the testing time. The data points on the left are taken from room temperature and the data points on the right are taken from the box. It can be seen that the water adsorption performance of sand is closely related to the relative humidity of the environment. With the increase of relative humidity, the water adsorption properties of sand increases.

It is worth mentioning that, due to large fluctuations of the relative humidity in the box, which is different from the natural environment where the sands normally stay, it will cause frequent water vapor exchange between sand surface and the environment, so the test results are different from water absorption performance of the sand in real environment, but the conclusions can still have some reference value.

3. Triboelectrificationmodel with adsorbed water particle system

Sand movement is a typical gas-solid two-phase flow movement, and the energy in the system is transferred from the gas phase to the sand phase. The friction energy caused by tangential displacement and the hysteresis deformation caused by the normal displacement can be converted into the internal energy of the sand to increase its temperature during the collision between the sand particles. At the same time, the temperature difference between the sand and the environment will enhance the heat exchange between the sand surface and the environment. Therefore, the air flow will not only affect the trajectory of sand particles, but also affect

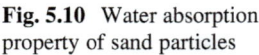

Fig. 5.10 Water absorption property of sand particles

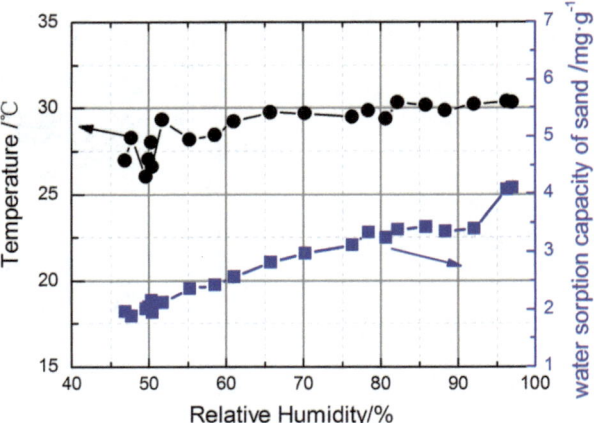

the temperature of the particle surface, thus affecting the process of sand friction charging.

It is assumed that the initial temperature of sand is the same as the ambient temperature, and the influence of normal and tangential deformation on the surface temperature is neglected. The change of sand temperature is related to the friction work and hysteresis deformation. The amount of internal energy W in the process of friction between sand i and sand j can be expressed as:

$$W = \mathbf{F}_{tij}^{C} \cdot \delta_{tij} + \mathbf{F}_{nij}^{C} \cdot \delta_{nij} \tag{5.32}$$

where, $\mathbf{F}_{tij}^{C}$ and $\mathbf{F}_{nij}^{C}$ are respectively the tangential and normal forces at the collision of the sand particles, δ_{tij} and δ_{nij} are the tangential displacement and the normal displacement at the collision of sand particles.

Assuming that the converted internal energy is distributed evenly between the two sand grains, the increase of temperature ΔT_1 of the particle i caused by collisional friction work can be expressed as

$$\Delta T_1 = \frac{a_1 W/2}{V_i \cdot \rho \cdot c} \tag{5.33}$$

Where,

V_i the volume of particle i;
ρ the density of particle i;
c the specific heat of particle i;
a_1 frictional heat transfer coefficient, in this chapter we take $a_1 = 0.9$.

In the process of sand friction, heat transfer between particles is neglected. Sand particles transfer the heat with its surrounding atmosphere in the wind field through convection and radiation, etc. According to the principle of heat balance, the internal heat change of sand with time should be equal to the change of heat in the surrounding environment through convection heat transfer, namely:

$$-\rho c V_i \frac{dT}{dt} = \alpha A (T - T_f) \qquad (5.34)$$

Where, α is the coefficient of convective heat transfer [W·m^{-2}·K^{-1}], T_f is the environment temperature, the initial condition is when $t = 0$:$T = T_0$
Integral to the above equation:

$$\frac{T - T_f}{T_0 - T_f} = e^{-\frac{\alpha A}{\rho c V}t} \qquad (5.35)$$

In the formula: $\Delta T_1 = T_0 - T_f$, that is the temperature increase of sand particle due to the friction colliso, let $\Delta T_i = T - T_f$, that is, the instantaneous temperature of particles.
For spherical sand particles: $V/A = d_s/6$.
Therefore, the convection conversion factor can be expressed as:

$$\alpha = -\frac{\rho c d_s}{6t} \ln \frac{\Delta T_i}{\Delta T_1} \qquad (5.36)$$

Then, the Nusselt number N can be expressed as:

$$Nu = \frac{\alpha d_s}{\lambda} = -\frac{\rho c d_s^2}{6\lambda t} \ln \frac{\Delta T_i}{\Delta T_1} \qquad (5.37)$$

where λ is the thermal conductivity [W·m^{-1}·K^{-1}].
The empirical formula of Nu for the heat transfer around the sphere is:

$$Nu = 0.37Re^{0.6}Pr^{1/3} \qquad (5.38)$$

The temperature is increased due to the collision of the sands, and the instantaneous temperature of the sand i after separation is:

$$\Delta T_i = e^{-\frac{Nu}{B}t} \Delta T_1 \qquad (5.39)$$

where, $B = \frac{\rho c d_s^2}{6\lambda}$.
From the discussion in the previous section, we can see that a thin layer of water molecules will be adsorbed on the sand surface due to the porous nature of the sand particles. If the ambient temperature is maintained, the concentration of H$^+$/OH$^-$ in the water molecules increases with the increase of the temperature in the sand surface. When two sand particles with temperature difference come into contact with each other, H$^+$/OH$^-$ will migrate in the direction of concentration difference.

The flow rate J of H^+/OH^- caused by the temperature difference between the sand particles i and sand j can be expressed as

$$J = a_2 a_3 \left(\Delta T_i - \Delta T_j \right) \tag{5.40}$$

where, ΔT_i and ΔT_j are the temperature differences between the sand particle i or the sand particle j and the environment, respectively. Hence $(\Delta T_i - \Delta T_j)$ indicates the temperature difference between the sand particle i and the sand grain j. a_2 is the dielectric constant of the sand surface with adsorbed water, which is related to relative humidity and ambient temperature in the sand system. a_3 is a proportionality coefficient indicating the potential difference between the particles caused by the temperature difference between the particles due to ion migration, which is related to the ion product of water and the migration speed of H^+/OH^-.

The dielectric constant a_2 of the sand surface increases with the increase of relative humidity, while the coefficient a_3 changes with the relative humidity. When the relative humidity is small, the increment of the relative temperature will cause a rapid increase of the ion/electron concentration on the sand surface. However, when the relative humidity exceeds a certain threshold value, the thickness of the adsorbed water thin layer on the sand surface becomes larger. The surface temperature of sand will increase or decrease due to the collision and at the same time form the temperature difference with the water vapor around the sand. The water vapor and sand surface will restrain the change of the particle surface temperature by heat transfer. The specific heat of water is five times that of the sand, meaning that the heat capacity of water is large, and the thermal inertia of the temperature change is large. When humidity is relatively large, due to the strong inhibition of heat exchange of water vapor, the temperature of sand surface is not easy to change, and the collision process cannot produce static electricity. If the relative temperature is below a certain value, a thin layer of water molecules on the sand surface will not be formed or will only be partially formed. At this point, there is a high degree of randomness for ion transport between the collided particles, meaning that the migration of ions occurs when there is a thin layer of water molecules on the contact surface, resulting in a sharp reduction in the total amount of sand charged or no static electricity. The dielectric constant of humid soil is between $0.005\ S\cdot m^{-1}$ and $0.45\ S\cdot m^{-1}$, and the value of dielectric constant increases with increase of soil water content. Since the dielectric constant of sand is not found in the literature, we take the dielectric constant $a_2 = 0.1\ S\cdot m^{-1}$ and assume $a_3 = 0.8$ in this paper as the reference of the dielectric constant of the soil.

Therefore, the charge amount of sand particles can be expressed as:

$$\begin{cases} q_i = \int_{t_s}^{t_e} J dt \\ q_j = -\int_{t_s}^{t_e} J dt \end{cases} \tag{5.41}$$

In the process of collision, the temperature of sand with large particle size is low while the small size is high. Because the migration speed of H+ is faster than that of

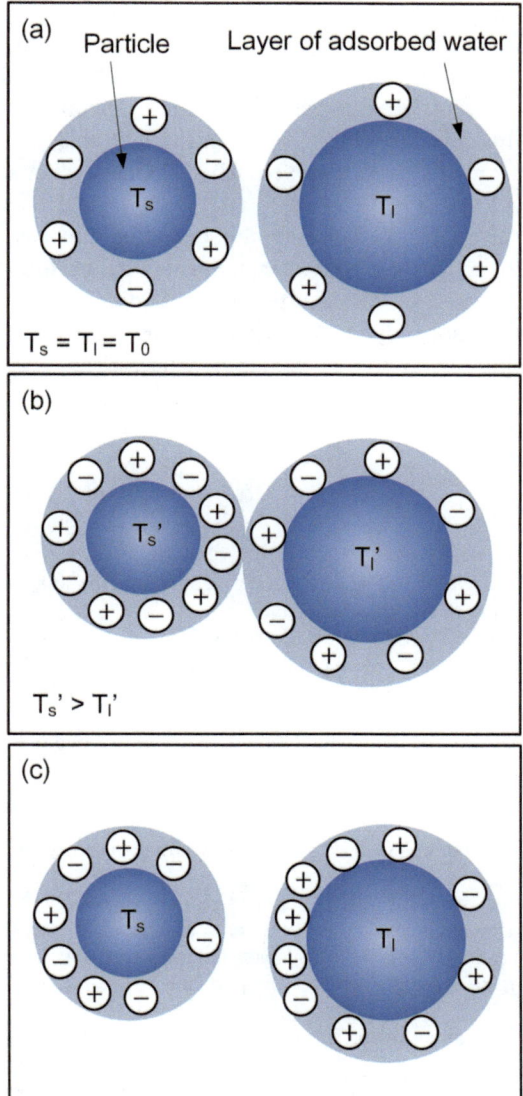

Fig. 5.11 Particle contact friction charging model with adsorbed water particle system. (**a**) Taking the particles with different particle sizes as an example, assume that the initial temperature of the particle surface is equal, i.e. $T_s = T_l = T0$, where, the subscript s represents particles with small particle size and l represents particles with large particle size. H+ and OH− ions are adsorbed on the surface of the particles; (**b**) Particles collide with each other under the action of air flow, and the mechanical energy in the process of friction is converted into the internal energy which causes the rise of particle temperature. Assuming that the internal energy is evenly distributed among the contact particles, the temperature rise of particles with small particle size is large and the temperature rise of the particles with large particle size is small due to the difference in the particle size. The concentration of ions on the absorbed water layer of the particle surface increases with the rise of particle temperature, and ions migrate towards the temperature difference; (**c**) Since the migration rate of H+ is faster than that of OH− ions, more H+ migrates from particles with relatively high temperature to particles with relatively low temperature; after the separation, particles with large particle size showed positive charge, and particles with small particle size showed negative charge. (Gu et al. 2013)

OH-ion, more H+ migrates from the sand particles with relatively high temperature to the sand particles with relatively low temperature. Therefore, in the contact process of two sands, H^+ migrates from the small size particles with higher temperature to the large size particles with lower temperature, and generally, the sands particles with small particle size are negatively charged, while sands with large particle size are positively charged. The schematic diagram of the physical model is shown in Fig. 5.11.

The particle surface formed during the industrial process is not absolutely smooth and there is a thin layer of water molecules on the surface. Therefore, the triboelectrification model of the adsorbed water particle system described in this section is applicable not only to the particle system present in nature, but also applicable to the particle system in industrial processes in, such as fluidized bed, pulverized coal transport and so on.

References

Desch, S.J., and J.N. Cuzzi. 2000. The generation of lightning in the solar nebula. *Icarus* 143 (1): 87–105.

Grzybowski, B.A., M. Fialkowski, and J.A. Wiles. 2005. Kinetics of contact electrification between metals and polymers. *The Journal of Physical Chemistry B* 109 (43): 20511–20515.

Gu, Z., W. Wei, J. Su, and C.W. Yu. 2013. The role of water content in triboelectric charging of wind-blown sand. *Scientific Reports* 3 (1377).

Hu, W., L. Xie, and X. Zheng. 2012. Simulation of the electrification of wind-blown sand. *The European Physical Journal. E, Soft Matter* 35 (3): 1–8.

Kok, J.F., and D.J. Lacks. 2009. Electrification of granular systems of identical insulators. *Physical Review E* 79 (5): 051304.

Kok, J.F., and N.O. Renno. 2008. Electrostatics in wind-blown sand. *Physical Review Letters* 100 (1): 014501.

Lacks, D.J., and A. Levandovsky. 2007. Effect of particle size distribution on the polarity of triboelectric charging in granular insulator systems. *Journal of Electrostatics* 65 (2): 107–112.

Lacks, D.J., N. Duff, and S.K. Kumar. 2008. Nonequilibrium accumulation of surface species and triboelectric charging in single component particulate systems. *Physical Review Letters* 100 (18): 188305.

Lowell, J. 1979. The relationship between contact charging and the concentration of donor impurities in polymers. *Journal of Physics D: Applied Physics* 12 (12): 2217–2222.

Lowell, J., and W.S. Truscott. 1986a. Triboelectrification of identical insulators. I. An experimental investigation. *Journal of Physics D: Applied Physics* 19 (7): 1273.

Lowell, J., and W.S. Truscott. 1986b. Triboelectrification of identical insulators. II. Theory and further experiments. *Journal of Physics D: Applied Physics* 19(7): 1281–1298.

Pahtz, T., H.J. Herrmann, and T. Shinbrot. 2010. Why do particle clouds generate electric charges? *Nature Physics* 6 (5): 364–368.

Xie, L., G. Li, N. Bao, and J. Zhou. 2013. Contact electrification by collision of homogenous particles. *Journal of Applied Physics* 113 (18): 184908.

Chapter 6
Numerical Modeling Methods for Particle Electrification

When the sand particles move in the flow field with the sand particles and the flow field coupling with each other, the fluid is not only affected by the internal force of the fluid, but also by the force of the sand on the fluid. Following is the CFD-DEM method used to simulate the gas-solid two-phase flow model.

6.1 Governing Equations of Gas-Solid Two-Phase Flow

6.1.1 Governing Equations of Fluid

From the perspective of numerical calculation, single-phase fluid motion is controlled by incompressible continuity equation and momentum equation. When the particles are added to the fluid, the fluid mesh is not entirely fluid, and the fluid is not only affected by the fluid internal force, but also by the force of particles on the fluid. When the particle concentration is large, the governing equation of the gas-solid two-phase flow can be obtained by introducing the porosity ε to modify the density, viscosity and source term of the fluid governing equation.

$$\frac{\partial(\varepsilon\rho_f)}{\partial t} + \nabla \cdot \left[(\varepsilon\rho_f)\mathbf{u}_f\right] = 0 \tag{6.1}$$

$$\frac{\partial\left[(\varepsilon\rho_f)\mathbf{u}_f\right]}{\partial t} + \nabla \cdot \left[(\varepsilon\rho_f)\mathbf{u}_f\mathbf{u}_f\right] = -\varepsilon\nabla p + \nabla \cdot (\varepsilon\tau_f) + (\varepsilon\rho_f)\mathbf{g} - \mathbf{f}_{drag}$$
$$- \mathbf{f}_{Mag} - \mathbf{f}_{Saff} \tag{6.2}$$

where, $\tau_f = \mu_f[(\nabla\mathbf{u}_f + \nabla\mathbf{u}_f^T) - \frac{2}{3}\nabla \cdot \mathbf{u}_f\mathbf{I}]$ is the fluid shear stress. In this paper, we use the Smagorinsky model to describe the turbulent motion of the fluid, ε is the porosity of the fluid and can be expressed as

© Springer Nature Singapore Pte Ltd. 2017   117
Z. Gu, W. Wei, *Electrification of Particulates in Industrial and Natural Multiphase flows*,
DOI 10.1007/978-981-10-3026-0_6

$$\varepsilon = 1 - \sum_{i=1}^{n} V_i / \Delta V \qquad (6.3)$$

where V_k is the volume of particles, and ΔV is the volume of the fluid grid.

$\mathbf{f}_{drag}$, $\mathbf{f}_{Mag}$, $\mathbf{f}_{Saff}$ is respectively the unit volume of drag force, Magnus force and Saffman force, their respective expressions are as follows:

$$\mathbf{f}_{drag} = \frac{1}{\Delta V} \sum_{i=1}^{n} \mathbf{F}_{drag,i} \qquad (6.4)$$

$$\mathbf{f}_{Mag} = \frac{1}{\Delta V} \sum_{i=1}^{n} \mathbf{F}_{Mag,i} \qquad (6.5)$$

$$\mathbf{f}_{Saff} = \frac{1}{\Delta V} \sum_{i=1}^{n} \mathbf{F}_{Saff,i} \qquad (6.6)$$

6.1.2 Governing Equation of Particle Motion

The motion of particle follows Newton's law of mechanics, and the governing equation of particle motion can be expressed as:

$$m_i \frac{dv_i}{dt} = \mathbf{F}_{drag,i} + \mathbf{F}_{Mag,i} + \mathbf{F}_{Saff,i} + m_i g + \sum_j \mathbf{F}_{ij}^C \qquad (6.7)$$

$$I_i \frac{d\omega_i}{dt} = \sum_j M_{ij} \qquad (6.8)$$

$$M_{ij} = \sum_j \left(r_s n_{ij} \times F_{tij}^C \right) \qquad (6.9)$$

where,

v_i	the linear velocity of particle i;
ω_i	the angular velocity of particle i;
F_{ij}^C	the contact force acted on the particle i when the particle i collides with the particle j;
M_{ij}	the contact torque acted on particle i when particle i collides with the wall surface;
$F_{drag,i}$	drag force due to viscous shearing stress;
$F_{Mag,i}$	the lift caused by the rotation of the particles due to the particle collisions with each other;
$F_{Saff,i}$	the lift caused by the rotation of the particles due to the velocity gradient of the fluid.

6.1.3 Force Between Particle and Fluid

When sand particles move in the air, gas and sand interaction can be described by interphase or drag force. Because of the complexity of sand-air interactions, the drag force between air and sand is usually calculated by using empirical or semi-empirical models. In this paper, the drag force model of Ergun/Wen and Yu is used.

$$F_{drag,i} = \frac{\beta(\mathbf{u}_f - \mathbf{v}_i)}{\rho} \tag{6.10}$$

the value of β depends on the porosity ε:

$$\beta = \begin{cases} \dfrac{1-\varepsilon}{\varepsilon^2 D_p}\left[150\dfrac{(1-\varepsilon)\mu_f}{D_p} + 1.75\rho_f\varepsilon|\mathbf{v}_i - \mathbf{u}_f|\right] & (\varepsilon \leq 0.8) \\ \dfrac{3}{4}C_D\dfrac{|\mathbf{v}_i - \mathbf{u}_f|\rho_f(1-\varepsilon)}{D_p}\varepsilon^{-2.65} & (\varepsilon > 0.8) \end{cases} \tag{6.11}$$

$$C_D = \begin{cases} 24(1 + 0.15Re^{0.687})/Re & (Re \leq 1000) \\ 0.43 & (Re > 1000) \end{cases} \tag{6.12}$$

$$Re = \frac{\rho_f D_p \varepsilon |\mathbf{v}_i - \mathbf{u}_f|}{\mu_f} \tag{6.13}$$

where,

$\mathbf{v}_i$ the velocity vector of the sand particles;
$\mathbf{u}_f$ the velocity vector of the fluid;
C_D drag coefficient;
Re Reynolds number of fluid;
D_p equivalent diameter of sand group

In fact, when the sand moves in the wind field, in addition to gravity and air drag force, it also rotates because of the collision and other causes, thus forming the lift perpendicular to the relative velocity of particles and air. The rotation speed of sand particles in the air flow can reach a few hundred to thousands of revolutions per second. Therefore, Magnus effect of sand movement in the wind field should not be ignored. The Magnus effect has a significant influence on the sand jump, with an increase of jump height about 20% (White and Schulz 1977).

$$F_{Mag,i} = \frac{\pi}{8}d_i^3\rho_f\left[(\boldsymbol{\omega}_f - \boldsymbol{\omega}_i) \times (\mathbf{u}_f - \mathbf{v}_i)\right] \tag{6.14}$$

where, $\boldsymbol{\omega}_f$ is the angular velocity of fluid movement, $\boldsymbol{\omega}_f = \frac{1}{2}\nabla \times \mathbf{u}_f$.

Zou et al. (2007) calculates the influence of Saffman force on the jump height and horizontal distance of sand is 4.6% and 3.7% respectively. Therefore, the influence of saffman force is also considered in the calculation.

$$\mathbf{F}_{saff,i} = 1.61d_i^2\left(\mu_f\rho_f\right)^{1/2}\left|\Omega_f\right|^{-1/2}\left[\left(\mathbf{u}_f - \mathbf{v}_i\right) \times \Omega_f\right] \tag{6.15}$$

where, Ω_f is the vorticity of the fluid, $\Omega_f = \nabla \times \mathbf{u}_f$.

6.1.4 The Contact Force Between Particles

In the DEM model, each particle is subjected to its own mass force and interparticle contact force. When the particle i contacts with a plurality of particles, the contact force of the surrounding contacting particles against the particle i is calculated. The contact force models are (Gu 2010): linear model, non-slip normal model based on Hertz theory (H-MDns), and nonlinear hysteresis model (H-MD). Compared the solutions of the three models with the exact solutions or the experimental results, Renzo and Maio concluded that the increase of the model complexity does not lead to a significant improvement in the calculation accuracy. If the parameters are rational, the error between simple linear models and exact solutions or the test results is very small. The linear model is briefly described below, and a detailed description of the inter-particle contact force is given by Gu (2010).

In fact, the contact between the two particles is a surface contact rather than a point contact. In the DEM calculation, the contact force can be decomposed into the normal contact force $\mathbf{F}_{n,ij}$ and the tangential contact force $\mathbf{F}_{t,ij}$, respectively, as shown in Fig. 6.1.

The model of the contact force among the particles and between the particle and the wall can be represented by a spring-damper-friction plate. When particles collide with other particles and the wall, the normal is subjected to both the elastic force and damping force, respectively described as the first and second of the equation on the right. According to the literature (Tsuji et al. 1993), when the particles contact with each other or the wall the normal stress relations are respectively:

$$\mathbf{F}_{n,ij} = -\left(\kappa_{n,i}\delta_{n,ij}\right)\mathbf{n}_i - \eta_{n,i}\left(\mathbf{v}_{ij} \cdot \mathbf{n}_i\right)\mathbf{n}_i \tag{6.16}$$

$$\mathbf{F}_{nw,ij} = -\left(\kappa_{nw,i}\delta_{n,ij}\right)\mathbf{n}_i - \eta_{nw,i}\left(\mathbf{v}_{ij} \cdot \mathbf{n}_i\right)\mathbf{n}_i \tag{6.17}$$

where,

$\kappa_{n,i}$ the elastic coefficient among particles in normal direction;
$\eta_{n,i}$ the damping coefficient among particles in normal direction;
$\delta_{n,ij}$ the displacement among particles in normal direction;
$\kappa_{nw,i}$ the elastic coefficient between particles and the wall in normal direction;
$\eta_{nw,i}$ the damping coefficient between particles and the wall in normal direction;

$\mathbf{n}_i$ and $\mathbf{t}_i$ represents the unit vectors of the particles i in the normal and tangential directions, respectively. The subscripts i and j denote the particle i and the particle j.

Fig. 6.1 Contact force of
particles collision

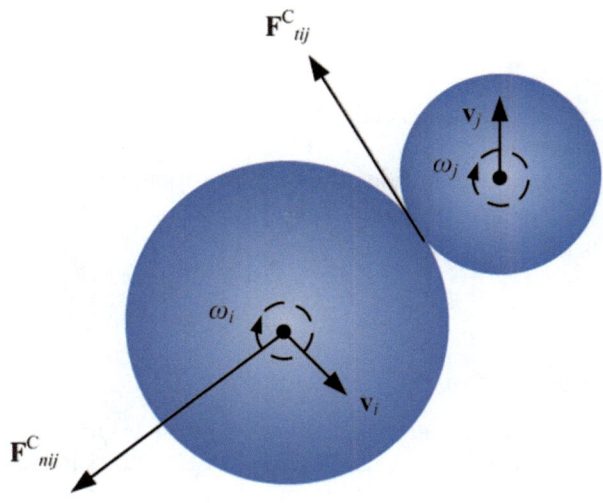

The particle-particle normal relative displacement can be calculated by the following formula:

$$\delta_n = \mathbf{R}_i + \mathbf{R}_j - \left| \mathbf{xyz}_j - \mathbf{xyz}_i \right| \tag{6.18}$$

where,

R   the radius vector of particles;
xyz   the coordinate vector of the particles.

If it is the particle-wall collision, the normal relative displacement δ_n is

$$\delta_n = \mathbf{R}_i - \left| (\mathbf{xyz}_i - \mathbf{xyz}_w) \cdot \mathbf{n} \right| \tag{6.19}$$

$\mathbf{v}_{ij}$ represents the relative velocity vector of particle i and particle j

$$\mathbf{v}_{ij} = \mathbf{v}_i - \mathbf{v}_j \tag{6.20}$$

For the particle-wall collision, because regarding the wall as stationary particles,
there are

$$\mathbf{v}_{ij} = \mathbf{v}_i \tag{6.21}$$

where $\mathbf{n}$ is the unit normal vector when particle i collides with particle j, and its
direction is from the centroid of particle i to the centroid of particle j, and the
normal unit vector is calculated as:

$$\mathbf{n} = \frac{\mathbf{xyz}_j - \mathbf{xyz}_i}{\left|\mathbf{xyz}_j - \mathbf{xyz}_i\right|} \tag{6.22}$$

For particle-wall collisions, $\mathbf{n}$ is perpendicular to the wall and toward the exterior of the wall.

Similar to the normal force calculation, for a particle-particle and particle-wall collision, the tangential force $\mathbf{F}_{t,ij}$ can be written as:

$$\mathbf{F}_{t,ij} = -\kappa_{t,i}\boldsymbol{\delta}_{t,ij} - \eta_{t,i}\mathbf{v}_{t,ij} \tag{6.23}$$

$$\mathbf{F}_{tw,ij} = -\kappa_{tw,i}\boldsymbol{\delta}_{t,ij} - \eta_{t,i}\mathbf{v}_{tw,ij} \tag{6.24}$$

where,

$\kappa_{t,i}$ the elastic coefficient among particles in tangential direction;
$\eta_{t,i}$ the damping coefficient among particles in tangential direction;
$\boldsymbol{\delta}_{t,ij}$ tangential displacement among particles;
$\kappa_{tw,i}$ the elastic coefficient between particles and the wall in tangential direction;
$\eta_{tw,i}$ the damping coefficient between particles and the wall in tangential direction;

$\mathbf{v}_{t,ij}$, $\mathbf{v}_{tw,ij}$ represent the sliding velocity of particle at the collision point, and for the collision of particle-particle, particle-wall, it can be expressed as

$$\mathbf{v}_{t,ij} = \mathbf{v}_{ij} - \left(\mathbf{v}_{ij} \cdot \mathbf{n}\right)\mathbf{n} + \left(L_i\boldsymbol{\omega}_i + L_j\boldsymbol{\omega}_j\right) \times \mathbf{n} \tag{6.25}$$

$$\mathbf{v}_{tw,ij} = \mathbf{v}_{ij} - \left(\mathbf{v}_{ij} \cdot \mathbf{n}\right)\mathbf{n} + L_i\boldsymbol{\omega}_i \times \mathbf{n} \tag{6.26}$$

$\boldsymbol{\delta}_{t,ij}$ is the tangential deformation of the particle generated by the tangential force, the expression is

$$\boldsymbol{\delta}_{t,ij} = -\mathbf{n} \times \left(\mathbf{n} \times \boldsymbol{\delta}_{t\text{-}\Delta t,ij}\right) + \mathbf{v}_{ij}\Delta t \tag{6.27}$$

$\boldsymbol{\omega}_i$, $\boldsymbol{\omega}_j$ represents the angular velocity vector of particle i and particle j respectively, L_i, L_j represents the actual arm of force length. Through the geometric relationship, we can get

$$L_i = \frac{\left(\left|\mathbf{xyz}_j - \mathbf{xyz}_i\right|^2 + R_i^2 - R_j^2\right)}{2\left|\mathbf{xyz}_j - \mathbf{xyz}_i\right|} \tag{6.28}$$

$$L_j = \left|\mathbf{xyz}_j - \mathbf{xyz}_i\right| - L_i \tag{6.29}$$

If it is particle-and-wall collision,

$$L_i = |(\mathbf{xyz}_i - \mathbf{xyz}_w) \cdot \mathbf{n}| \tag{6.30}$$

Since the amount of grain deformation is very small, L_i, R_j can be approximately replaced by the particle radius R_i, R_j.

If the particles are subjected to a tangential force greater than the friction force, that is

$$|\mathbf{F}_{t,ij}| > f|\mathbf{F}_{n,ij}| \tag{6.31}$$

The particles will slide, and at this time the tangential force of the particles can be given by the sliding friction formula:

$$\mathbf{F}_{t,ij} = f|\mathbf{F}_{n,ij}|\mathbf{t} \tag{6.32}$$

where, $\mathbf{t} = \frac{\delta_t}{|\delta_t|}$.

At the same time, there will be a number of particles colliding with particle i, and the force acted on the particle i is the resultant forces of all particles, so for the particle i, the resultant force and torque due to the collision is

$$\mathbf{F}_i = \sum_j \left(\mathbf{F}_{n,ij} + \mathbf{F}_{t,ij}\right) \tag{6.33}$$

$$\mathbf{T}_i = \sum_j \left(\mathbf{T}_{n,ij} + \mathbf{T}_{t,ij}\right) \tag{6.34}$$

In the model of particle collisions, there are three parameters including elastic modulus κ, damping coefficient η and friction coefficient f. These three parameters can be constants or can be determined by the physical properties of the particles: Young's modulus E_s and the Poisson's ratio σ_s according to the Hertz theory (Cundall and Strack 1979). For collision with the wall surface, the Young's modulus E_w and the Poisson's σ_w of the wall material are used.

According to Hertz's elastic collision theory, when the two particles collide, if the particles have the same radius and physical properties, the normal elastic coefficient is

$$\kappa_n = \frac{\sqrt{2R_{eq}}E_s}{3\left(1 - \sigma_s^2\right)} \tag{6.35}$$

For a particles-and-wall collision, there are

$$\kappa_n = \frac{4\sqrt{R_i}}{\sqrt{\frac{(1-\sigma_s^2)}{E_s} + \frac{(1-\sigma_w^2)}{E_w}}} \tag{6.36}$$

When the two particles collide, their tangential elasticity is

$$\kappa_t = \frac{2\sqrt{2R_{eq}}H_s}{2-\sigma_s}\delta_n^{1/2} \tag{6.37}$$

For particles-and-wall collision, there are

$$\kappa_t = \frac{8\sqrt{R_i}H_s}{2-\sigma_s}\delta_n^{1/2} \tag{6.38}$$

where, $H_s = \frac{E_s}{2(1+\sigma_s)}$.

Tsuji et al. (1993) has proposed a method for calculating the damping coefficient η, which can be written in the following form according to the Hertz elastic collision theory for a single-degree-of-freedom oscillation system in which a proton is connected to the wall by a spring and a damper.

$$m\frac{d^2x}{dt^2} + kx^{3/2} + \eta\frac{dx}{dt} = 0 \tag{6.39}$$

The recovery factor e is defined as the ratio of the velocity after the particle collision to the velocity before the collision, that is: $e = -\frac{v}{v_0}$.

If the physical properties of a particle are given, then the recovery factor should be a constant, independent of particle mass, elastic coefficient, and damping coefficient. Tsuji et al. (1993) points out that when the damping coefficient is written in the following form

$$\eta = a(mk)^{1/2}x^{1/4} \tag{6.40}$$

It is possible to satisfy the requirement that the recovery coefficient is constant. As the recovery coefficient is a commonly used physical parameter, it can be determined by experiment or looking up the literature. After determining the recovery factor e, the value of α can be determined by Fig. 6.2.

Therefore, the normal and tangential damping coefficients during particle collision are

$$\eta_n = a_n(mk_n)^{1/2}\delta_n^{1/4} \tag{6.41}$$

$$\eta_t = a_t(mk_t)^{1/2}\delta_t^{1/4} \tag{6.42}$$

Fig. 6.2 The relation between the dimensionless coefficient α and the restitution coefficient e (Tsuji et al. 1993)

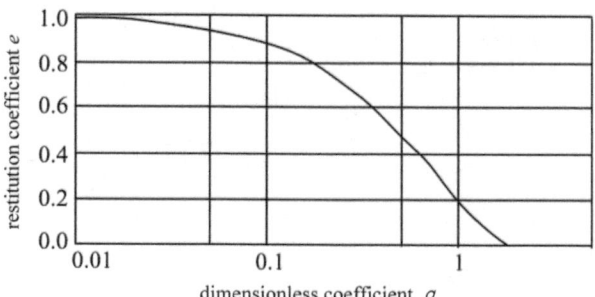

6.2 Applications of Numerical Simulation on Particle Charging

6.2.1 Charged Particle Movement in Electrostatic Precipitator

Electrostatic precipitators are used to collect suspended particles in gases using an electrostatic force and they are the one ways to control air pollution caused by industrial plants. The configuration mostly used in electrostatic precipitation technique is the wires-to-plates. It consists of high-field parallel active wires located midway between the grounded plates (the collecting electrodes) where the air flows through. The ions produced by the corona discharge near the wires charge the dust particles which are thus driven toward the collecting plates. The particle charges are neutralised and the particle is thus collected. The collection efficiency of the wire-to-plate electrostatic precipitators (WPESPs) depends on numerous variables like the global drift velocity of charged particles to be removed and their distributions, the magnitude of applied voltage, the active electrodes radius, the humidity and temperature of the air, etc. The basic corona discharge physics is well-known and it can be described as a self-sustaining electrical gas discharge occurring at the vicinity of high-field electrodes. In the WPESPs the high-field wires are surrounded by ionization region where the free charges are produced and a lowfield drift region where charged particles drift to the collecting plates. The corona drift region is governed by the Poisson's equation and the current continuity equation (Nouri and Zebboudj 2010).

Single wire line pipe type electrostatic precipitator will be taken as an example to study the law for the movement of the fluid and the trajectories of the particles with different particle diameters in the electrostatic precipitator, and inspect the factors influencing the efficiency of the electrostatic precipitator. The computational geometry model is shown in Fig. 6.3. Its geometry size is $L \times W \times H = 0.5\ m \times 0.1\ m \times 0.1\ m$, the diameter of the poles is $D = 1\ mm$, and the total number of meshes is 150,000, with an intensified mesh around the pole.

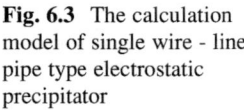

Fig. 6.3 The calculation
model of single wire - line
pipe type electrostatic
precipitator

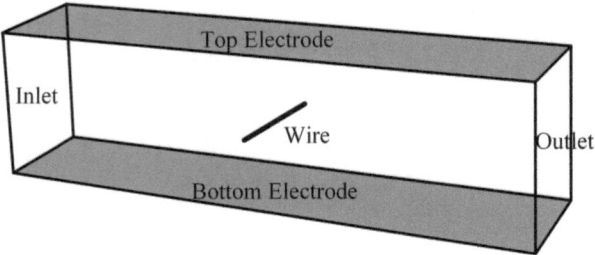

Boundary conditions are set as shown in Table 6.1. The boundary conditions of velocity, pressure, and voltage are common for the first or second boundary conditions. It is worth noting that the charge density boundary conditions on the surface of the poles are set using the above iterative method.

① Flow field structure

Figure 6.4 shows the flow line of the fluid flow between the plates at different inlet velocities at 30 kV, and it can be seen that when the velocity of the inlet fluid is small, the electric field has a significant effect on the motion of the fluid. When the velocity of the inlet fluid increases slowly, the influence of the electric field on the motion of the fluid gradually decreases. The black line in the center of Fig. 6.4 is the location of charged wire. When the air inlet velocity $u = 0.2$ m·s^{-1}, similar to the last section, around the charged wire will form four vortices which can take the air in the middle of the plate to the plate wall; When the air inlet velocity $u = 0.5$ m·s^{-1}, the vorticity gets smaller in the upstream direction of the charged wire, and the position of the vortex is closer to the conductor in the axial direction. Due to the blocking action of the charged wire, a downstream vortex appears, similar to the flow around the cylinder; When the inlet velocity increases to $u = 1.0$ m·s^{-1}, all the vortices disappear and the electric field has little effect on the air flow.

In practical industrial applications, if the air flow inside the electrostatic precipitator is not evenly distributed, it will have a great impact on the dust removal efficiency of the precipitator. For a particular type of electrostatic precipitator, its efficiency is limited to a range of dust. If the air inlet flow is greater than the maximum flow electrostatic precipitator set, the dust removal efficiency of precipitator will decline. Because of the increase of air velocity, the chance of combining the particles with the ionized gas ions will also reduce. Meanwhile, it will take away part of the particles deposited on the dust collector plate, which increases the secondary dust effect. From the principle of electrostatic precipitator, the lower the gas flow rate is, the greater the chance for the particles to charge in the flow will be, and the higher the dust removal efficiency will be.

② The particle trajectory

In order to clarify the movement of charged particles inside the electrostatic precipitator, a small amount of particles are incorporated into the inlet airflow,

Table 6.1 Electrostatic precipitator model boundary condition settings

	Velocity	Pressure	Voltage	Charge density
Inlet	Velocity inlet	Zero gradient	Zero gradient	Zero gradient
Outlet	Zero gradient	Constant pressure	Zero gradient	Zero gradient
Top plate	Sliding	Zero gradient	0	Zero gradient
Bottom plate	Sliding	Zero gradient	0	Zero gradient
Polar line	Sliding	Zero gradient	Set value	Peek Theorem

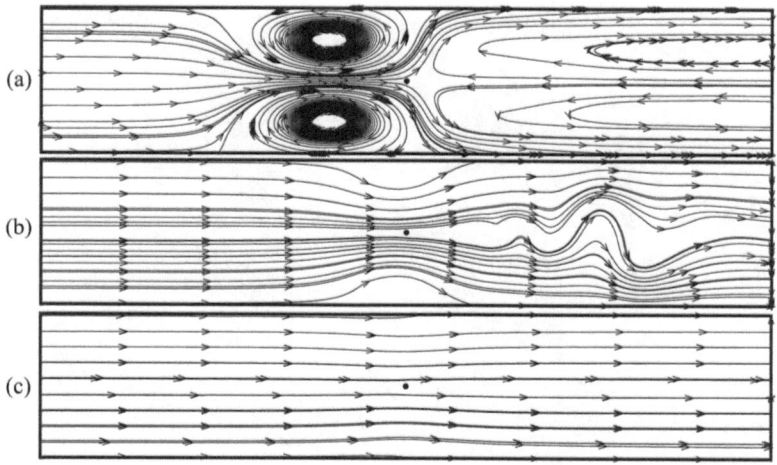

Fig. 6.4 Fluid flow between plates at different inlet velocities. Voltage: 30 kV (**a**) $u = 0.2$ m·s^{-1}; (**b**) $u = 0.5$ m·s^{-1}; (**c**) $u = 1.0$ m·s^{-1}

which are collectively referred to as dust-laden airflow, to simulate the trajectory of particles. Figure 6.5 shows the trajectories of particles in the dust-laden airflow inside the electrostatic precipitator in the flow field shown in Fig. 6.4. The color in the figure indicates the size of particles, ranging from 1 μm to 30 μm, exhibiting a normal distribution. It can be seen in Fig. 6.5 that particles with large particle sizes are apparently more dominated by the electric field force due to their large charge capacity, and are therefore collected and deposited on the plate first, while particles with smaller particle size are difficult to collect. When the inlet velocity of the dust-laden airflow is small (Fig. 6.5a, $u = 0.2$ m·s^{-1}), the particles stay in the electrostatic precipitator for a long time. Under the carrying of the upstream vortex of the charged wire, all the particles can be collected on the plate before they reach the position of the charged wire. When the inlet velocity of the dust-laden gas flow is large (as shown in Fig. 6.5b, $u = 0.5$ m·s^{-1}), a small amount of particles with smaller particle sizes can escape from the electrostatic precipitator. When the inlet velocity of the dust-laden air flow continues to be increased (Fig. 6.5c, $u = 1.0$ m·s^{-1}), the drag force increases and the effect of the electric force will be weaker, then there will be more particles with larger particles sizes escaping from the outlet, and unable to be collected, thus affecting the efficiency of electrostatic precipitators.

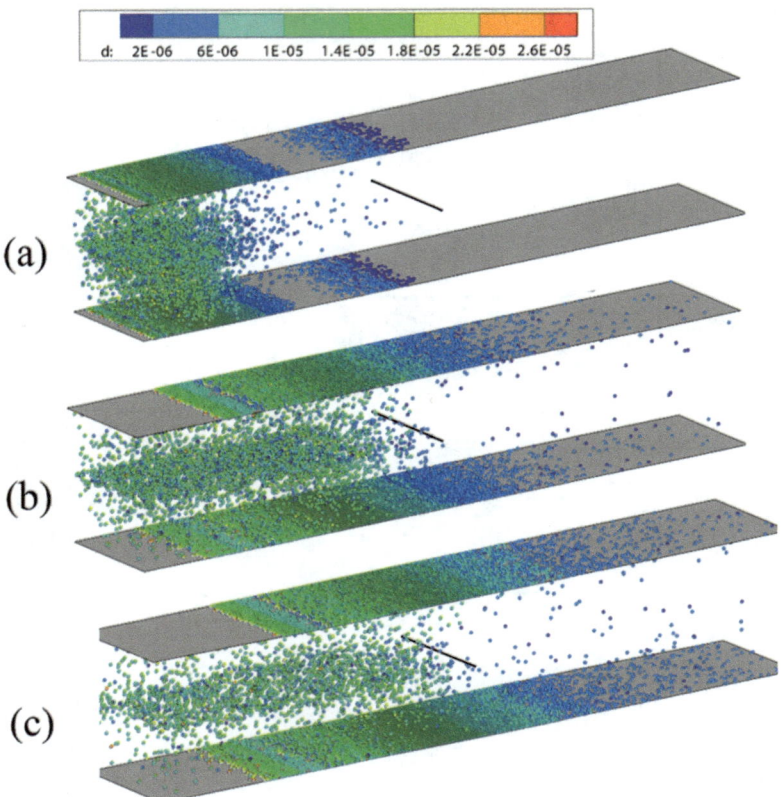

Fig. 6.5 Particles trajectory between the plate at different inlet velocity, t = 5s. Voltage: 30 kV (a) $u = 0.2$ m·s^{-1}; (b) $u = 0.5$ m·s^{-1}; (c) $u = 1.0$ m·s^{-1}

In order to reflect the trajectories of the particles in the electrostatic precipitator more clearly, the motion vector of the particles is shown in Fig. 6.7 when the line voltage is 30 kV and the velocity $u = 1.0$ m·s^{-1}. The color in the Figure represents the different velocities of particles. It is worth noting that the particles deposited on the plates remain in contact with the plates because of the programmed set-up, and the particles deposited on the plates remain at the same speed before they contact with the plate. Therefore, the particles deposited on the plates still have velocity vectors as shown in Fig. 6.6. From Fig. 6.6, it can be seen that all particles move towards the plate due to the influence of the electric field force on the upstream of the charged wire. Most of the particles are deposited at the cross section of the charged wire. Due to the large electric field near the charged wire, the velocities of particles deposited near the cross section of the charged wire are large, but there are still a small number of particles with small particle sizes around the charged wire moving towards the outlet, and these particles are difficult to collect.

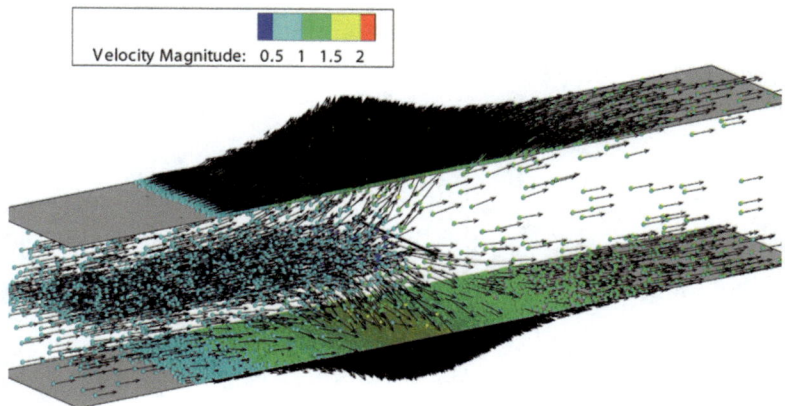

Fig. 6.6 Particle velocity vector in the electrostatic precipitator when $u = 1.0$ m·s^{-1}

③ The dust collection efficiency of electrostatic precipitator

Dust collection efficiency refers to the percentage of pollutants removed by electrostatic precipitator at the same time as the percentage of the pollutants entering the device, which is the main technical indicator to measure the performance of the dedusting device. Dust collection efficiency means the purification degree of the dust collection device can achieve to the dust-laden gas. In theory, the factors that affect the efficiency of electrostatic precipitator include the following aspects: the electrochemical properties of dust, gas temperature, humidity, flow rate and flow velocity of the dust air, dust concentration, uniformity of air distribution and the structure of dust collector itself.

It is clear that if the composition of the dust-containing gas to be treated and the structure of the precipitator itself are determined, the most intuitive factors affecting the efficiency of the electrostatic precipitator are the flow of the dust-containing gas and the voltage applied to the charged wire. The efficiency of electrostatic precipitator will be discussed in the two aspects.

In the previous section, the trajectories of particle movements at different inlet velocities of dust-laden gas are discussed under certain voltage conditions. In this section, the number of particles entering the electrostatic precipitator and the number of particles escaping from the electrostatic precipitator are calculated and the result of the efficiency of electrostatic precipitator at different inlet flow rates is shown in Fig. 6.7. At a certain control voltage, the dedusting efficiency of electrostatic precipitator decreases with the increase of inlet flow rate. When the inlet flow rate is small, the dust removal efficiency can reach 100%, but the treatment capacity of dust-laden gas is the smallest. When the inlet flow rate increases slowly, the dust removal efficiency decreases drastically, and the treatment capacity becomes weaker. Therefore, it is necessary to select the appropriate inlet flow rate according to the specific situation.

As is shown in Fig. 6.8, at a certain flow rate, under different operating voltages, dust removal efficiency of the electrostatic precipitator increases as the voltage increases and the electric field force on the particles increases, resulting in particles

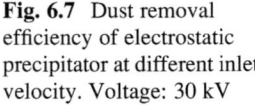

Fig. 6.7 Dust removal efficiency of electrostatic precipitator at different inlet velocity. Voltage: 30 kV

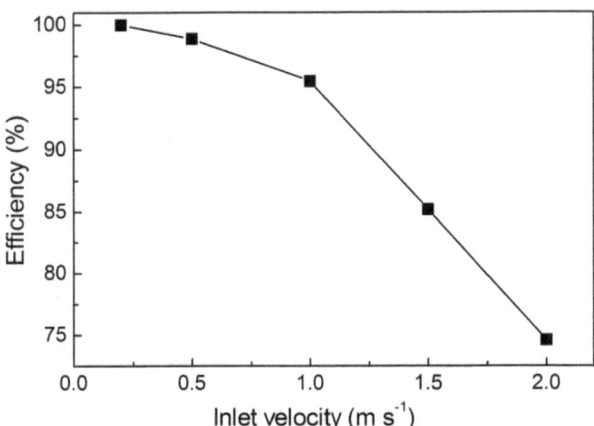

easily captured and deposited in the plate. But an exceedingly high operating voltage will have a negative impact on the safety of equipments and operating procedures, so a reasonable choice of operating voltage is also essential.

6.2.2   Charging Behavior of Sand in Horizontal Wind Tunnel

In this section, the whole process of wind-blown sand simulated, including the movement from start-up, transition to stabilization. The sand particles move under the action of wind field, and the sand and the wind field are exchanged for momentum, so that the sand accelerates the movement. The sand particles near the ground collide with each other and some sand particles are rebounded or splashing more sand particles. The sand particles that collide in the near-ground state will be charged before separation until the calculated charge-to-mass ratio tends to be stable, then sand flow is considered stable. Firstly, the initial and boundary conditions of the flow field (wind field) are set. The initial distribution and position of the sand are set. The instantaneous wind field can be obtained through solving the equation and updating the position of particles. If the particles collide, the carried charge of contacted sand can be calculated by using mentioned Sect. 5.2 of the adsorbed water and particle system friction charged model, and finally get the movement characteristics and charged characteristics of the sand system.

In the DEM calculation, the calculation of the time step is related to whether the mechanical behavior of particle collision can be reflected correctly, and directly influences the calculated charge of particles. From intuitive understanding, the calculated time step should be smaller than the particle contact collision time, so as to capture the particle collision behavior. However, if the calculated time step is too small, it will take a lot of calculation time, therefore the time step must be

Fig. 6.8 Dust removal efficiency of the electrostatic precipitator at different operating voltage. Inlet rate $u = 1.0$ m·s^{-1}

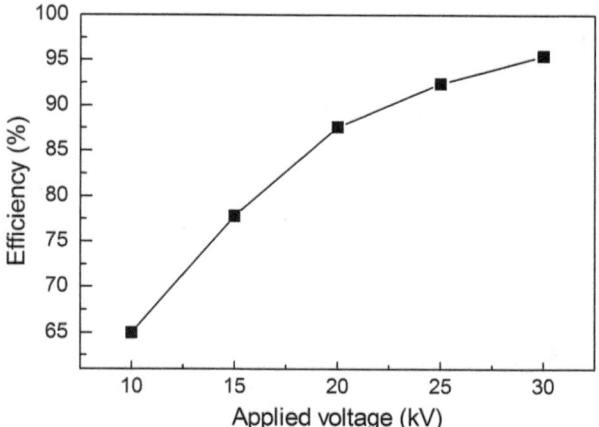

chosen rationally. Cundall (Cundall and Strack 1979) proposed that in order to keep a stable critical time step $\Delta t_c = \sqrt{m/\kappa}$, the time step chosen in this paper should be less than Δt_c which is 10^{-6} s.

Using the field wind tunnel of Shapotou Desert Experimental Station of Lanzhou Desert Research Institute, Chinese Academy of Sciences, Zheng et al. (2003) studied the saltation features of "uniform sand" (particle size within a certain range) and "mixed sand" (sand size range 40 ~ 600 µm), and measured their electric quantities. The wind tunnel is an indoor and outdoor DC closed suction-type active wind tunnel. The simulated sand flow in the wind tunnel is consistent with that in the field. The test section of the wind tunnel has 21 m for its lengths and 1.2 m for its width. The wind tunnel is made of aluminum alloy, which is grounded when measuring. The average electric charge per kilogram of moving sand was measured by measuring the weight of sand and the amount of electric charge from the inlet of the sand analyzer at the inlet of the sand collecting chamber, using a 20 cm high and 2 cm wide step sand collector. For the measurement of the electric field intensity at different heights in the sand flow field, a DPD-type atmospheric average electric field instrument was used. A more detailed description of the experiment is given in the literature (Zheng et al. 2003).

The number of sand particles in the calculation is limited due to the calculation conditions. Therefore, in order to increase the probability of collision among sand particles, the three-dimensional calculation model is adopted. The computational domain is slightly different from that of the wind tunnel. It has one tenth in terms of length and width and same height relative to the experimental wind tunnel, and uses the same inlet wind speed conditions. The calculation model diagram is shown in Fig. 6.9.

The main physical parameters of the simulation process are shown in Table 6.2. Where the wind profile of the inlet is given by the following equation:

Fig. 6.9 Calculation
domain of sand
electrification in wind
tunnel

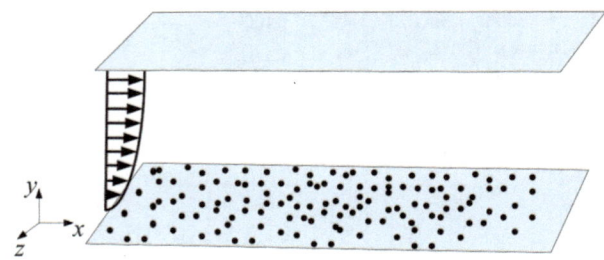

Table 6.2 The main physical parameters of the simulation process

Gas phase (air)		Solid phase (sand)	
Viscosity/m²·s⁻¹	1.46×10^{-5}	Particle size distribution	Single particle size and normal distribution
Density/kg·m⁻³	1.05	Density/kg·m⁻³	2650
Inlet wind speed/ m·s⁻¹	10,15,20	Shape	Spherical
Computational domain/m	$2.1(L) \times 0.12$ $(W) \times 1.2(H)$	Quantity	1 5000
Specific heat/ kJ·kg⁻¹·K⁻¹	1.006	Specific heat/ kJ·kg⁻¹·K⁻¹	3507

$$u(y) = \frac{u_*}{\kappa} \ln\left(\frac{y}{y_0}\right) \qquad (6.43)$$

Where $u(y)$ is the wind speed at height y, u_* is the friction velocity of the flow field, $\kappa = 0.4$ is the Karman constant, and $y_0 = D_p/30$ is the roughness of the bed (Anderson and Haff 1988). The relation between the friction velocity and the wind speed U is (Zhou et al. 2002):

$$u_* = \frac{U - 4.32}{11.50} \qquad (6.44)$$

During the calculation, the sand particles are allowed to settle freely from the upper part of the computational domain, and the initial distribution of sand particles on the ground surface is obtained. Figure 6.10 shows the free-sedimentation process of sand particles with mixed particle size. The different colors represent different particle sizes. Under the action of gravity, sand moved downward, collided with the ground, bounced, and finally when the energy slowly depleted, deposited on the ground.

After the initial position distribution of the sand particles is obtained and the initial wind speed is given at the inlet, the sand particles move under the action of the wind field. The calculation domain is shown in Fig. 6.9. It is worth mentioning that, because of the limited number of sand particles simulated, when the sand collided with the bottom with no sand, or the ground, it was assumed that sand

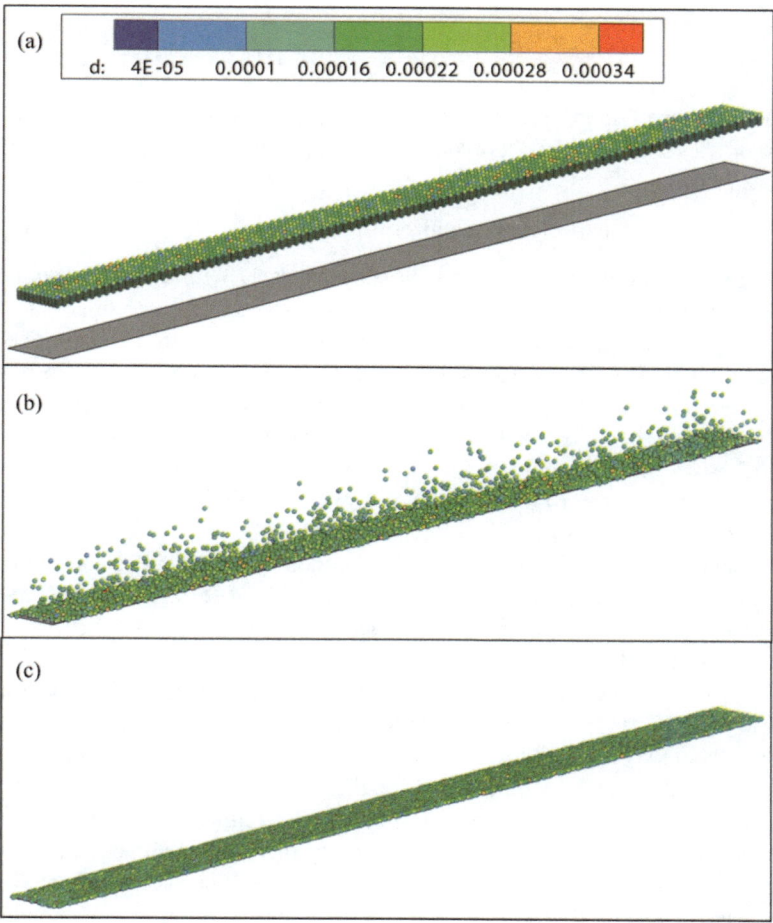

Fig. 6.10 Distribution of free-settling mixed sand at various time, the color indicates the particle size of sand (**a**) t = 0 s; (**b**) t = 0.5 s; (**c**) t = 10 s

particles collided with a particle size of 1000 μm, in order to estimate the collision effect between sand and sand bed; to set the periodic boundary of sand particles, meaning when sands are blown out of the computational domain, the velocity and the charged quantity are initialized and then put back to the inlet of the computational domain, giving the following saltate velocity and angle (Nalpanis et al. 1993):

$$f(u_s) = \frac{1}{\sqrt{2\pi} \cdot 2u_*} e^{-\frac{(u_s - 4u_*)^2}{8u_*^2}} \tag{6.45}$$

$$f(\alpha_s) = \frac{1}{\sqrt{2\pi} \cdot 19} e^{-\frac{(\alpha_s - 38)^2}{2 \times 19^2}} \qquad\qquad (6.46)$$

The number of sand particles and the amount of charges are calculated at the outlet of the computational domain to calculate the charge-mass ratio of sand particles.

① Charges of sand particles with uniform particle sizes

Figure 6.11 shows the movement of sand particles with a particle diameter of $D = 200\ \mu m$ in the wind tunnel with an inlet wind speed of $U = 15\ m \cdot s^{-1}$. The color of the graph indicates the amount of charges that the sand particles carry. The initial sand bed, obtained by free settling, is directly started up under the influence of inlet wind speed. Once the sand particles separated from the ground surface and entered into the air flow, it can be easily accelerated by the wind to obtain momentum from the airflow. A portion of the sand particles will have sufficient momentum to rebound when they return to the ground or to cause other surface particles to start up into the airflow due to impact effect of particle.

The sand particles with equal size are collided symmetrically, and the internal energy is evenly equal distributed into the two sand particles. After each collision, the temperature rises of both sand particles are the same, but because of the turbulence of the atmosphere, the probability of sand particles collision is different, with high sand temperature in high probability of collision and low temperature in the low probability of collision, which causes the relative temperature difference between sands. When the two sand particles with temperature difference come into contact, the charged ions migrate through the thin layer of water film. After separation, the sand particles are opposite charged. Figure 6.12 shows the temperature rise and the amount of charges carried by the sand particles with a uniform particle size of $D = 200\ \mu m$ after the friction collision. It can be seen from Fig. 6.12 that the amount of sand charge depends on the temperature difference between the sand particles that contact with each other. The charge of sand particles with the relative high temperature is not necessarily the largest; only when there is temperature difference between sands, the charges would migrate. Due to the randomness of the collision process, the charge of the sand particle with high temperature may be positive or negative.

② Charge of sand particles with a mixed particle sizes

As shown in Fig. 6.13, the charge-to-mass ratio of sand particles with different particle sizes at different inlet heights and the wind speed of $20\ m \cdot s^{-1}$ varies with time. After the initiation of sand movements for a period of time, with the onset of the collision process and the generation of temperature difference, the sands began to charge and the charge-to-mass ratio increased sharply. As the sand flow stabilized, the collision frequency became stable, and the charge-to-mass ratio became stable as well. Due to the randomness of the collision process, the change of charge-to-mass ratio is generally fluctuating.

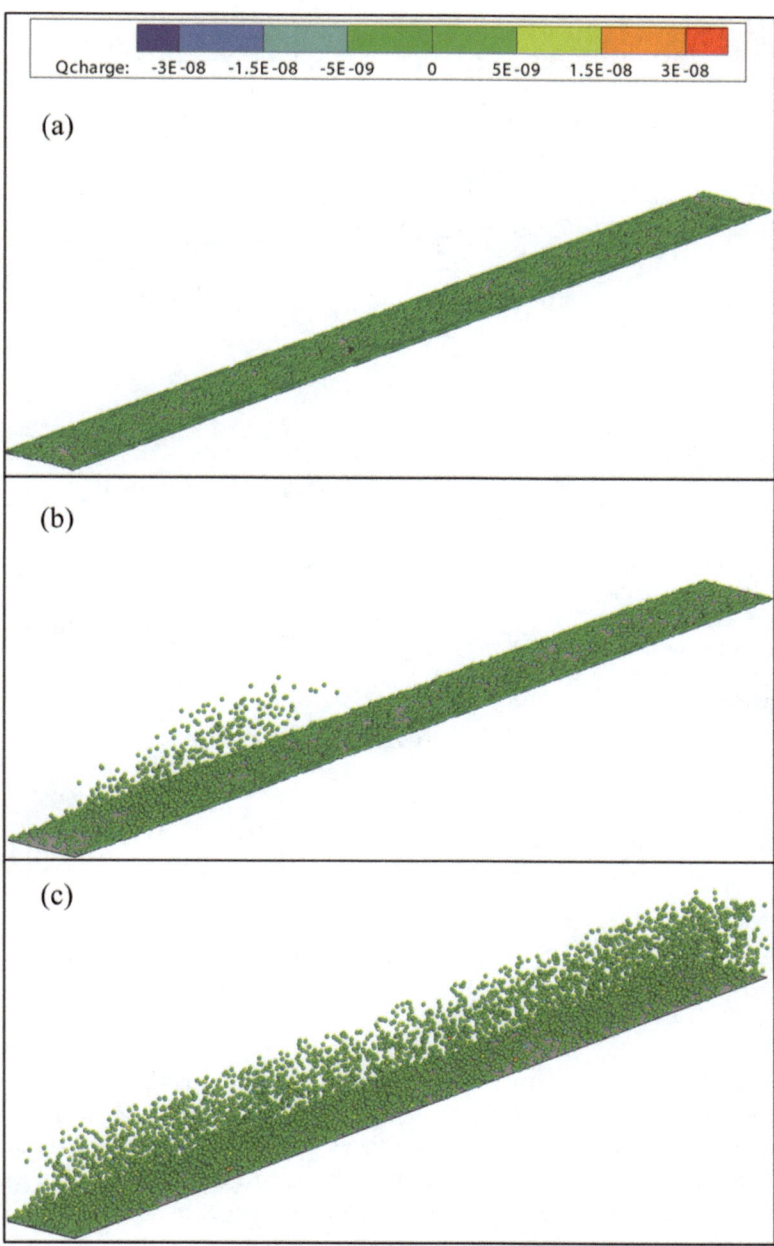

Fig. 6.11 The trajectory of the sand with uniform particle size in the wind tunnel, color indicates the charge, unit: μC. (**a**) t = 0 s; (**b**) t = 0.1 s; (**c**) t = 50 s

Fig. 6.12 The relationship between the temperature difference and the amount of charge

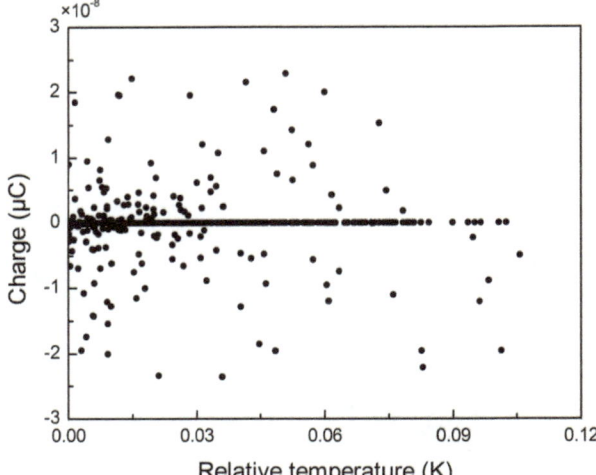

Fig. 6.13 The charge-to-mass ratio of particles at different height varies with time, the small picture shows the particle size distribution of sand particles

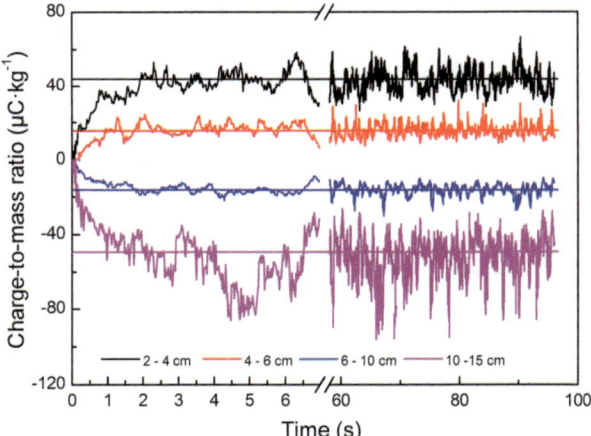

As the kinetic energy of sand movement comes from the potential energy transformation of sand particles, the temperature of sand particles increases with the non-symmetrical collision of sand particles with non-uniform particle sizes, and the increase of temperature is related to the sand particle sizes. The general tendency of particle temperature increase is that temperature of small size is high, and that of large size is low. When the sand particles are separated from each other at different temperatures, more H^+ ions migrate from the relatively high temperature sand particles (small sand particles) to the relatively low temperature sand particles (large sand particles). The results are: The large sand particles are positively charged and the small sand particles are negatively charged. Therefore, the general trend of the charge is that large sand particles are positively charged, and small sand particles are negatively charged, as is shown in Fig. 6.14. According to

Fig. 6.14 The charge of sand particles with different particle size

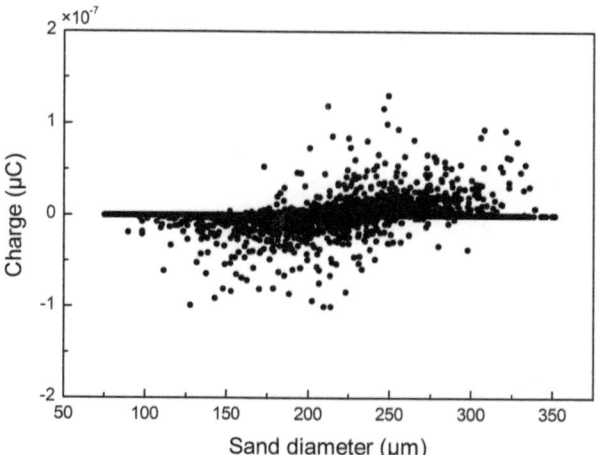

the simulation results, when the particle size is larger than 275 μm, the sand is positively charged; when the particle size is less than 175 μm, the sand is negatively charged. It is worth mentioning that the critical size of positive/negative charge of sand particles is affected by the diameter, velocity and relative humidity of the sand.

Figure 6.15 compares the mean particle diameters of different heights at different inlet wind speeds. It can be seen that in the case of low wind speed, the particle size of sand particles on the bed surface is larger, and with the increase of the height, the average particle size of sand in the low wind speed is larger than that of high wind speed at the same height. In other words, the greater the wind speed is, the greater energy it brings to the sand system, and the bigger the sand particle size can be blown up by the wind.

Figure 6.16 compares the numerically simulated results of the average charge-mass ratio of sand particles at different height with the measured results of Zheng et al. (2003) at two different inlet wind speeds. It can be seen that the average charge of the sand particles increases with the height, and the small sand particles are more likely to be brought into the air due to the relatively larger drag force of small sand particles, and the height of small sand particles are higher than the large sand particles. After the collision of sand particles of different sizes, the small sand particles are negatively charged and the large sand particles are positively charged. The reason for the discrepancy may be that the sand is assumed to be spherical in the simulation, and in fact the sand present in nature is essentially irregular. The contact area of the irregularly shaped sand particle is different from that of the ideal spherical sand particle during the collision, which will result in different calculated frictional forces, and eventually affect the amount of charge transfer.

③ Electrostatic field generated by charged sand

As collisions between the sands and between the sand and the ground make the sand charged, sand movement will generate an electrostatic field. According to

Fig. 6.15 The variation of mean particle size of sand with height under different wind speeds

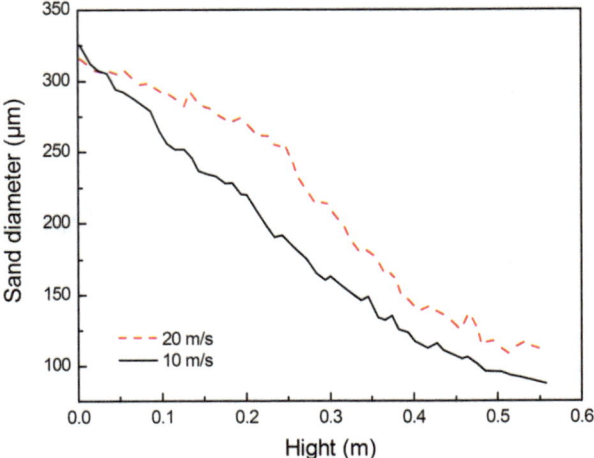

Fig. 6.16 The charge-to-mass ratio of sand particles calculated at different heights and is compared with the measured results (Zheng et al. 2003)

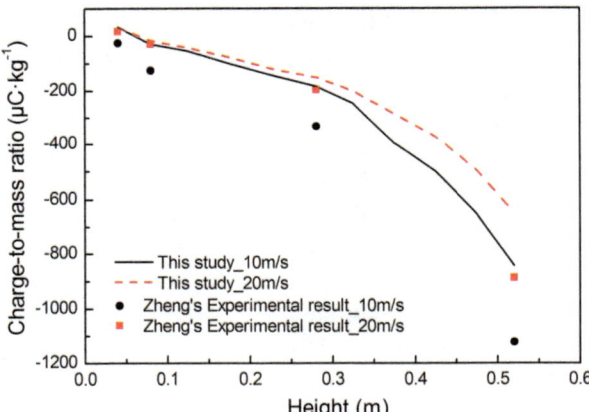

Coulomb's Law, the electric field generated at the position P (x, y, z) from the sand whose charge is q_i is

$$\mathbf{E}_i = \frac{q_i}{4\pi\varepsilon_0 r^3}\mathbf{r} \tag{6.47}$$

where, the dielectric constant in vacuum $\varepsilon_0 = 8.8542 \times 10^{-12} \, \mathrm{C^2 \cdot N^{-1} \cdot m^{-2}}$.

According to the superposition principle of electric field, the electric field generated at P (x, y, z) is

$$\mathbf{E} = \sum_{i=1}^{N} \mathbf{E}_i \tag{6.48}$$

Fig. 6.17 The electrostatic field strength generated by sand particles with uniform particle size

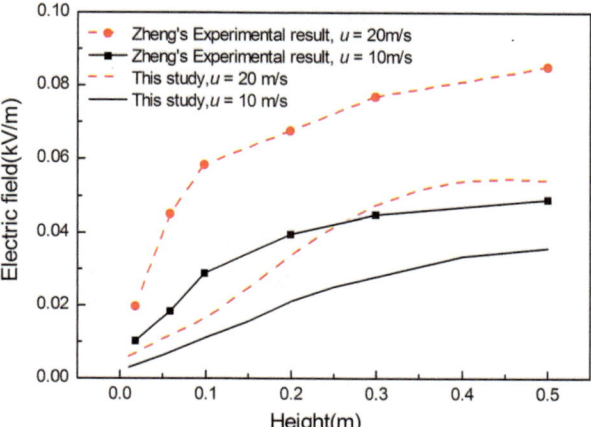

The electric field distribution along the height direction at 0.3 m away from the outlet is obtained after collecting all charged sand particles. Figures 6.17 and 6.18 show the contrast between the electric field measured and calculated in this paper respectively. It can be seen that the calculated results in this paper can basically reflect the change of electrostatic field generated in the sand movement, but it is smaller than the measured value, which may be due to the different probability of collisions between sand particles in the course of calculation, and because of turbulence in the flow field, the collision between sand particles is random.

In fact, the electric field strength shown in Figs. 6.17 and 6.18 is not calculated directly by the superposition of the electric fields generated by all the charged sand particles in the calculation, but is the result of statistical comparison. As described above, due to the limitations of the calculation conditions, this paper only simulated the movement characteristics and charging phenomenon of a limited number of sand particles in the wind tunnel, which is far from the actual number of sands in the wind tunnel. Taking the inlet wind speed $U = 20$ m·s^{-1} in Fig. 6.18 as an example, the sediment transport rate at the outlet height of 0.3 m is 0.06 kg·m^{-2}·s^{-1} in this paper, while the transport rate in the experiment is 1.5 kg·m^{-2}·s^{-1}(Zhou et al. 2002). Therefore, the electric field intensity directly obtained from the calculation does not reflect the electric field strength in the experiment. It is considered that the collision probability of sand particles in the calculation and experiment are the same, and the calculated electric field intensity is proportional to the ratio of sediment transport rate. This may also be one of the reasons for the deviation of the calculated results from the experimental results.

It is worth mentioning that the electric field strength is a parameter to characterize the macroscopic behavior of charged sand particles, which related to the number of sand particles, while the charge-to-mass ratio of sand particles is the characteristics of a single sand particle, which is a statistical average. Therefore, if we adopt the periodic boundary in this paper and the calculation time is long enough, we can get the true charge-to-mass ratio.

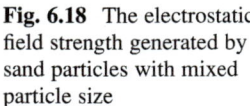

Fig. 6.18 The electrostatic field strength generated by sand particles with mixed particle size

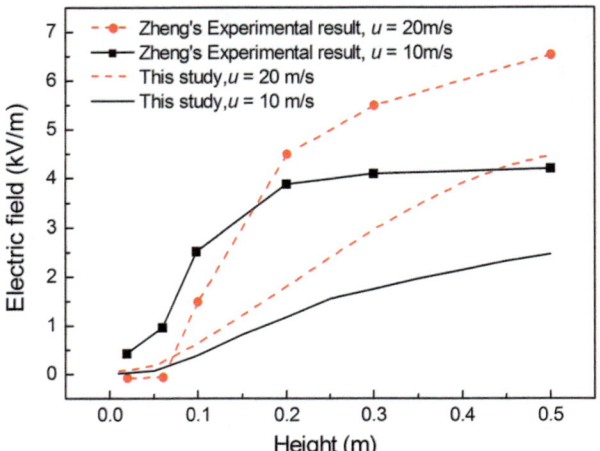

The electrostatic force has an influence on the movement of sand particles. From the visual point of view, the electric field force will change the trajectory of the charged sand particle. Therefore, the electrostatic field has a prominent impact on the time, the wind velocity profile and the sediment transport rate required by the sand flow from the start to reaching the equilibrium. Huang Ning et al.[34] pointed out that when the average sand charge is 60 $\mu C \cdot kg^{-1}$, the calculated single-width sand transport rate and sediment transport rate along the height distribution are in good agreement with the experimental results, which indicates that it is necessary to consider the action of the electrostatic force in the study of sand movement. When the electric field of sand and the average charge of sand particles are large enough, ignoring the electrostatic force may be one of the reasons that results of the previous sand movement model is different from the experimental result quantitatively, so it is necessary to consider the role of electrostatic force.

References

Anderson, R.S., and P.K. Haff. 1988. Simulation of Eolian saltation. *Science* 241 (4867): 820–823.

Cundall, P.A., and O.D.L. Strack. 1979. A discrete numerical model for granular assemblies. *Géotechnique* 29 (1): 47–65.

Gu, Z. 2010. *Wind-blown dust: Near surface gas-solid two-phase turbulent flow (in Chinese).* Beijing: Science Press.

Nalpanis, P., J.C.R. Hunt, and C.F. Barrett. 1993. Saltating particles over flat beds. *Journal of Fluid Mechanics* 251: 661–685.

Nouri, H., and Y. Zebboudj. 2010. Analysis of positive corona in wire-to-plate electrostatic precipitator. *European Physical Journal-applied Physics* 49 (1): 11001.

Tsuji, Y., T. Kawaguchi, and T. Tanaka. 1993. Discrete particle simulation of two-dimensional fluidized bed. *Powder Technology* 77 (1): 79–87.

White, B.R., and J.C. Schulz. 1977. Magnus effect in saltation. *Journal of Fluid Mechanics* 81 (03): 497–512.

Zheng, X., N. Huang, and Y. Zhou. 2003. Laboratory measurement of electrification of wind-blown sands and simulation of its effect on sand saltation movement. *Journal of Geophysical Research* 108 (D10): 4322–4331.

Zhou, Y.H., X. Guo, and X.J. Zheng. 2002. Experimental measurement of wind-sand flux and sand transport for naturally mixed sands. *Physical Review E* 66 (2): 021305.

Zou, X.Y., H. Cheng, C.L. Zhang, and Y.Z. Zhao. 2007. Effects of the Magnus and Saffman forces on the saltation trajectories of sand grain. *Geomorphology* 90 (1–2): 11–22.

Chapter 7
Experimental Methods for Particulate Charging Processes

Generally speaking, the phenomenon of electrostatic electrification is remarkable on products of polymer materials such as rubber, plastic, chemical fiber, etc. In light of common range of definition, the volume resistivity of electrical insulator is generally above 10^{11} Ω·cm. However, materials with volume resistivity from 10^{11} Ω·cm to 10^{15} Ω·cm fall into the range of materials that most easily generates charges, thus materials involved in electrostatic measurement always have a very high insulation resistance, some even reaching $10^{18} \sim 10^{20}$ Ω. For these objects with high resistance, it is very easy to accumulate static charges because electric charges are difficult to leak.

Learning from the foregoing description, the generation of static is related together with instant behavior (such as contact, separation, friction, stripping) between objects, so the "power supply" with electrostatic electrification is far more different the condition of galvanic electricity which can continuously supply objects with a large amount of electric charges. Therefore, under the condition of electrostatic electrification, the carried charges and stored electrostatic energy by objects are always very small, and generally, the static current is pico-ampere level or micro-ampere level. Because of the small electric quantity of static, even if a small amount of charges flow from charged body into test instrumentation or leak will cause great measuring error.

Electrostatic measurement is more significantly and remarkably affected by environmental conditions than common strong current or weak current measurement. Not only the environment temperature and humidity have great influence on the electrification of object and measuring result, but also the electrification status, shape and dimension and inter-allocation of other objects surrounding the measured object have complex influence on electrostatic parameters of measured object. Meanwhile, at different places and under different conditions, external energy, like pressure, light and electromagnetic field and so on will also impact the measuring result with varying degrees.

In general, surface resistance of insulator will reduce with rising temperature and humidity, due to the reason that the increasing of migration rate of impurity ions

© Springer Nature Singapore Pte Ltd. 2017

Z. Gu, W. Wei, *Electrification of Particulates in Industrial and Natural Multiphase flows*, DOI 10.1007/978-981-10-3026-0_7 and the effect of surface adsorption of water accelerate the leaking speed of charges. Wherein, the influence by humidity is far more obvious than that by temperature. Therefore, when taking electrostatic measurement, the variation of air humidity will certainly cause variation of related parameters. So during the process of measuring, the variation of temperature and humidity conditions should be taken into careful consideration. Except for making record well for analysis, the conditions of temperature and humidity should be controlled when there is a high requirement. Consequently, when taking measurements under laboratory conditions, related standards always specify to take measurement under a lower relative humidity (RH) which is, in general, ranging from 20 to 40%, while the temperature always ranges from 20–25 °C.

When taking measurement for electrostatic parameters, many environmental mechanics conditions will also affect the measuring result of electrification status of the measured object. Wherein, the effect of air pressure and mutual friction status cannot be neglected.

The experiment indicates that the charges carried by objects are relevant to ambient air pressure to some degree. Under a certain air pressure, the charges carried by objects show a minimal value. The occurrence of spark discharge has a closer relationship with air pressure because the ambient air pressure during discharging gap influences the impact velocity and motion free path of gas ions. Figure 7.1 shows the relationship between spark discharging voltage and the product of air pressure and discharging gap length, and it can be seen that the discharging voltage will present a minimal value when adopt a certain product. For different gases, this kind of relationship is similar, which is called the Paschen Curve.

The friction status between objects directly affects the charges carried by objects. The stable value is mainly determined by characteristics of the object itself (such as work function), the surface state and so on. When two objects are making mutual friction, the relationship between the charges carried by object and friction velocity can be given by the formula $Q = f(v)A^{1/2}$, wherein A represents frictional work. This formula indicates that the carried charges Q is a function of friction velocity. The experiment indicates that carried charges Q will be increased with increasing friction velocity v in a certain velocity range (when $v < 0.1$ m·s^{-1}).

Actually, in various electrostatic measurements, it often occurs that the measured results of the same electrostatic parameter of the same object may have great difference when taking repeated measurements, and the results will often have greater differences if the interval of each measurement is longer, meaning that the reproducibility of electrostatic measurement is usually worse than other measurements (strong current or weak current). In essence, the lower level of reproducibility of electrostatic measurement is attributed to and determined by the complexity of electrostatic electrification phenomenon.

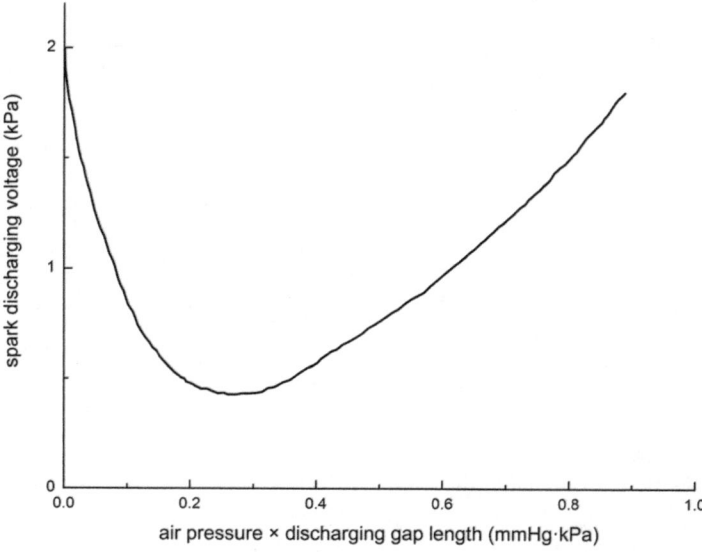

Fig. 7.1 Relationship between spark discharging voltage and the product of air pressure and discharging gap length

7.1 Measurement of Electrostatic Potential (Pressure)

Electrostatic voltage is the difference between the electrostatic potential of some point on charged body's surface and the potential of a specified reference point (usually the "ground"). Because the potential of the ground is generally adopted as zero, the electrostatic potential value of charged body's surface represents its electrostatic pressure level. Since potential is a physical quantity in proportion to charge, the potential relatively reflects the charged degree of object, meaning that the measurement of pressure (potential) can be used to recognize the amount of carried charges.

Electrostatic pressure is usually measured by electrostatic voltmeter which includes contact making electrostatic voltmeter and non-contact electrostatic voltmeter.

7.1.1 Method Using Contact Making Voltmeter

Contact making electrostatic voltmeter is electrified through the measuring probe contacting the measured object and works by utilizing electrostatic force action principle. This kind of instruments have two stationary and inter-insulated metal box electrodes A and B (electrode A is connected to probe and electrode B is grounded), and a rotatable metal sheet C hung on the metal wire with a small

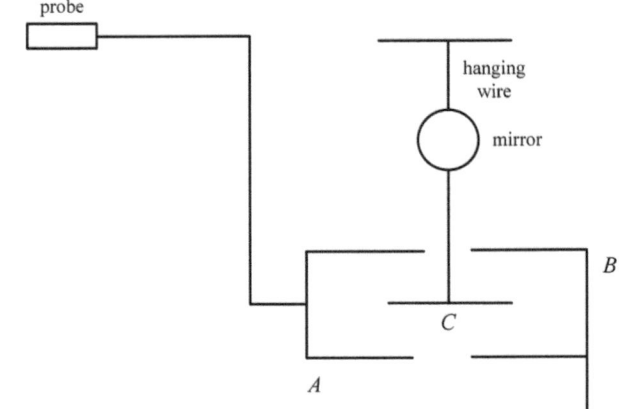

mirror attached. Schematic diagrams of instrument structure are as shown in
Fig. 7.2.

When the measuring probe in Fig. 7.2 contacts the charged object, electrode A is
electrified due to contacting and forms electric field between electrode A and B. In
this electric field, metal sheet C is electrified due to electrostatic induction and
deflected by the action of electric field force of electrode A and B, leading to the
deflection of both the hanging wire and the small mirror. When the deflecting torque
is equilibrated with the reaction torque of hanging wire, the deflection angle
represents the measured voltage level. The voltage represented by the angle can
be shown by the small mirror fixed on the wire through cursor.

Contact making voltmeter is easy to use with simple and firm structure, and its
price is low. However, this kind of voltmeter cannot distinguish the polarity of
measured potential and is only suitable for measuring voltage on conductor. The
measuring error is relatively small and the accuracy error of measurement can be
less than 2%. When an insulated object is electrified, contact making voltmeter
cannot be used to prepare measurement because that the molecules in insulator
strongly constrain free electrons, resulting in inconformity of potential distribution
at each point on the surface, and the electrified state of object usually changes after
contacting measuring probe. Moreover, this kind of instrument is inconvenient to
use for occasions that the measuring probe cannot make contact.

7.1.2 Method of Using Non-contact Voltmeter

Non-contact voltmeter makes use of electrostatic induction or air ionization principles to work, and the voltmeter does not have to contact a charged body during
measurement. This kind of non-contact voltmeter includes rotating blade type,
vibrating capacitance type, direct induction type and collecting type. Electrostatic
induction type instrument is to place the probe near the charged body and obtain the

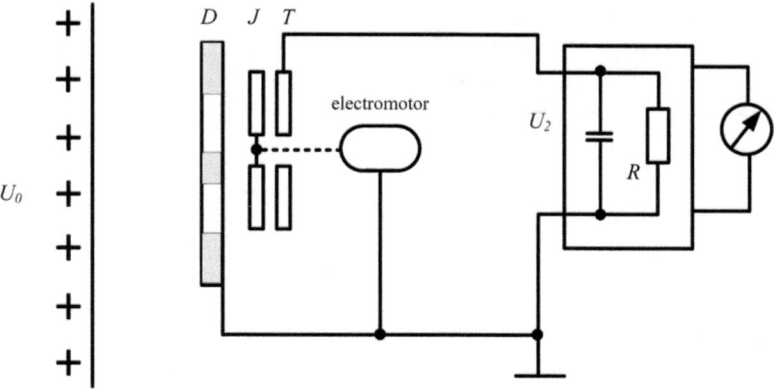

Fig. 7.3 Schematic of rotating blade type instrument

data of the surface potential by measuring the surface electric field of the charged body. The air ionization type instrument is to utilize radioisotope to ionize air, and measure the voltage to ground of charged body in accordance with the resistance voltage divider.

The accuracy of non-contact voltmeter measurement is superior to 15%. All the non-contact voltmeter can be fabricated into portable type with small size and light weight so as to be used for field measurement.

① Rotating blade type instrument

Rotating blade type instrument belongs to the AC amplification instrument of electrostatic induction type, and is comprised of measuring probe, amplifier and display unit. The principle of this structure is as shown in Fig. 7.3.

T in the Fig. 7.3 is probe (pole piece) of the instrument, two metal electrodes D and J are inserted between the T and the measured charged body, wherein the D is stationary and grounded with cross-shaped opening on it. The J is driven by micro motor and can rotate (i.e. rotating blade), and the shape of J is exactly the same as the shape of opening of electrode D. The pole piece T has a central opening through which the rotating shaft connecting pole piece J and the motor passes without contact.

When the pole piece J is rotated by the motor, the aperture of the grounding electrode D is periodically covered or opened, so that the receiving probe pole piece T periodically senses the electrostatic voltage signal of the measured object. Thus, the induced signal on the T is modulated into AC signal, and then shown by the gauge outfit through impedance transformation, AC amplification, and detection.

In order to identify the polarity of charged body, a power supply with known polarity may be connected to the metal sheet J as a reference power supply. At this point, when there is no external induced signal, the probe T will produce AC induced voltage and instrument pointer will be deflected as long as the instrument is turned on, with the position of the pointer serving as the zero position. Where the polarity of the charged body is the same as the polarity of the rotating blade, the pointer is deflected to the positive direction, and vice versa.

According to theoretical calculations, the voltage $U(t)$ at the instrument input resistance R is given by the following formula:

$$U(t) = U_0 \frac{n\omega C_0}{\sqrt{\left(\frac{1}{R}\right)^2 + n^2\omega^2 C_0^2}} \cos \omega t \qquad (7.1)$$

Wherein:

U_0 electrostatic of the measured object, V;
C_0 capacitance between the measured object and electrode D, F;
n the number of blade;
ω angular frequency of the rotating blade, rad·s^{-1};
C input capacitance of the instrument, F.

As can be seen from Eq. (7.1), the voltage signal obtained at the input resistance R of the instrument periodically changes with the time t. For this reason, such instruments can be used for long-term or continuous monitoring of electrostatic environment, and are widely used with advantages of good stability, high sensitivity, and relatively simple structure.

② Vibrating capacitance type instrument

Vibrating capacitance type instrument is an AC amplification instrument of electrostatic induction type, which utilizes the mechanical vibration of metal electrodes to modulate the induced signal into AC signal. The working principle of vibrating capacitance type instrument is as shown in Fig. 7.4.

The position of probe T in the figure is not stationary, but periodically vibrates along the horizontal direction in the figure, hence the capacitance C_0 between the probe T and the measured charged body is periodically changed, and:

$$C_0 = C_\infty(1 + K \sin \omega t) \qquad (7.2)$$

Wherein:

C_∞ static capacitance between the probe T and the measured object at equilibrium position, F;
ω angular frequency of the mechanical vibration of the probe, rad/s;
K modulation factor of C_0, as the amplitude of probe is very small, $K \ll 1$;

According to the theoretical derivation, the voltage signal at instrument input resistance R can be obtained as:

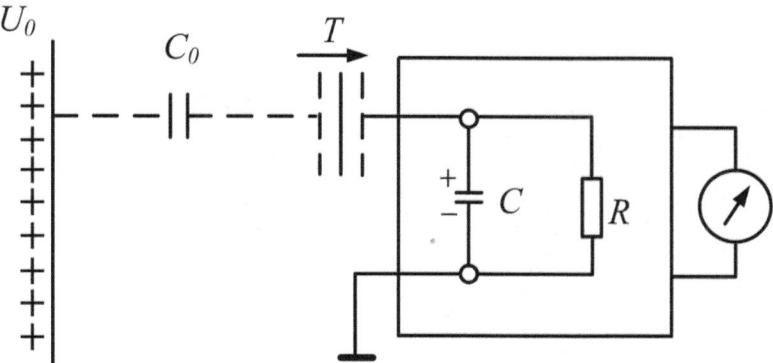

Fig. 7.4 Schematic of vibrating capacitance type instrument

$$U(t) = U_0 \frac{\omega C_0 K}{\sqrt{\left(\frac{1}{R}\right)^2 + \omega^2 C^2}} \cos \omega t \tag{7.3}$$

Wherein:

U_0 electrostatic potential of the measured object, V;
C input capacitance of the instrument, F;
R input resistance of the instrument, Ω;

The meanings of other symbols are the same as in Eq. (7.2).

As the measured signal is very weak, the impedance converter adopting high impedance input is used to receive signal. After inputting the signal after impedance conversion to the range converter, the signal is shown by the gauge outfit through amplification by AC amplifier and detection.

As seen from the Eq. (7.3), the voltage $U(t)$ at the instrument input resistance R periodically changes with the time t, therefore such instruments with good stability and high sensitivity can be used for long-term and continuous potential monitoring. The disadvantage of such instruments is the relatively complex structure of dynamic capacitor, which resulting in the relatively complex structure of the instrument.

③ Direct induction type instrument

The working principle of direct induction type instrument is as shown in Fig. 7.5.

In the Fig. 7.5, A is the measured charged body, T the measuring probe; C_0 between A and T representing the capacitance between the measured object and probe, C_1 the probe to ground capacitance, R and C the input resistance and capacitance of the instrument respectively.

Assuming that the actual electrostatic voltage of A to ground is U_0, then the electrostatic voltage U will be induced on T due to the existence of C_0, and a part of the induced charges will leak out to ground through R. Thus, the measured voltage will be:

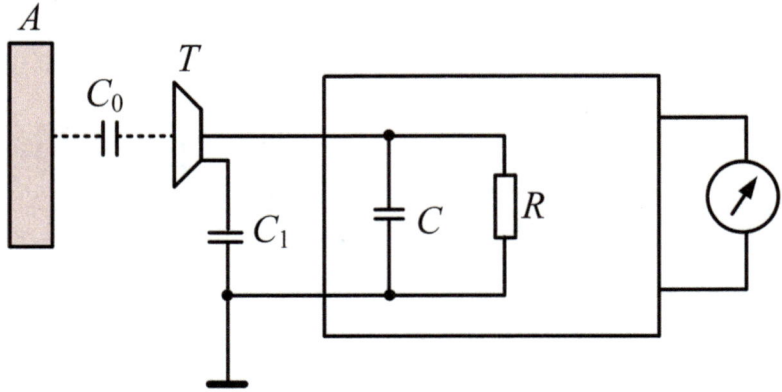

Fig. 7.5 Schematic of direct induction type instrument

$$U(t) = \frac{C_0}{C_0 + C} U_0 e^{-\frac{t}{RC}} \tag{7.4}$$

It can be seen that the induced voltage $U(t)$ on the probe attenuates with the measuring time t. When the position of the instrument and the measured charged body is relatively fixed, C_0 fixed. The $\frac{C_0}{C_0+C}$ in Eq. (7.4) may be regarded as a constant, the voltage U_0 on the measured object can be obtained through the induced voltage U measured by the probe. Obviously, the ratio between U and U_0 can be changed (i.e. the range of instrument can be changed) by changing the relative position of the probe and the measured object (i.e. changing C_0).

As seen from the Eq. (7.4), the measured voltage U is related to the time t. If the t is longer, there will be more charges leaking from R, hence the U will be smaller, meaning that the error with actual measured voltage U_0 will be greater. In order to increase the discharge time constant τ (requiring $\tau > 180$ s), the input resistance R of the instrument is required to be large enough.

Obviously, such instruments are not suitable for long-term monitoring. As the induced signal is weak, there is a need to set various DC amplifiers in actual measuring mechanisms to amplify the signal. Such kind of instrument has simple structure, small size, light weight, and a portable performance, but it also has low sensitivity and poor stability.

④ Collecting type instrument

Collecting type instrument belongs to the ionized non-contact instrument. The test working principle and equivalent circuits are as shown in Fig. 7.6.

In the Fig. 7.6, measuring probe is composed of a lead metallic shielding cylinder (grounding), a radioactive source F (usually adopt radioactive isotope such as Ra 226), and collector plate with porous J. Fill in polytetrafluoroethylene material between the collector J and the shielding metallic cylinder to fix and insulate them.

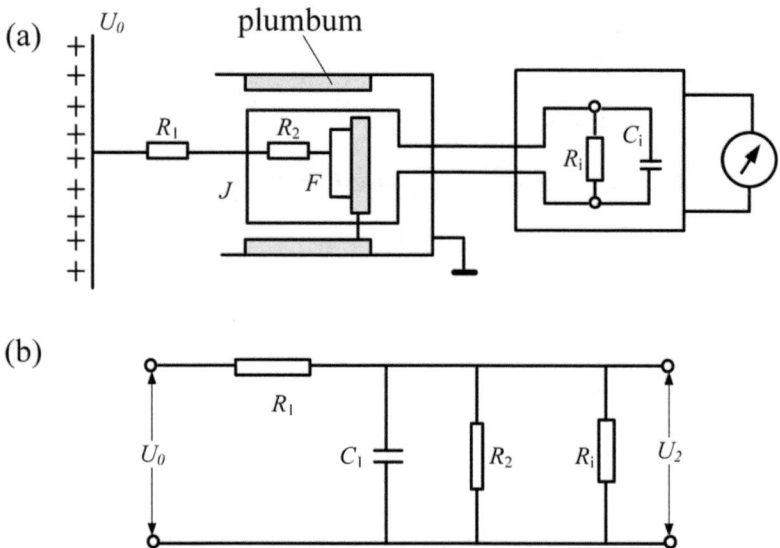

Fig. 7.6 (a) Schematic of collecting type instrument and (b) equivalent circuit

The α and β rays emanated from the radioactive source are emitted through the small window to ionize the surrounding air, and then open a stable ionization region near the collector small window. When the probe is introduced into the vicinity of the charged body, the resulting electric field will move ions with a particular sign (as same as the sign of the charged body) in the ionization region to the collector, thus forming a directional ionic current. The direction and magnitude of the ionic current are determined by the polarity and intensity of the charged body. Ionic current signal is read by gauge outfit through DC amplifying.

Setting that the measured object's electrostatic potential is U_0, the equivalent resistance between the object and the measuring probe is R_1, and the resistance between the collector and the ground is R_2, from the equivalent circuit diagram, the voltage U_2 on the collector is:

$$U_2 = \frac{R_2}{R_1 + R_2} U_0 \tag{7.5}$$

Apparently, for a specific instrument, R_2 is a constant value, so the change of difference between the measured voltage U_2 and the actual potential U_0 of charged body can be achieved by changing R_1 (changing the distance between the measured object and the probe), thus achieving the purpose for changing the instrument range.

When taking measurement, collecting type instrument is less affected by the external electric field, but the air ions generated by air ionization caused by radioactive rays and a portion of charges of charged body will achieve recombination action, so that the actual potential of charged body is reduced, resulting in measurement errors. Therefore, it is inappropriate to prolong the measurement time. Furthermore, if there is wind in test phenomena or if the charged body moves too fast, the number of ions in the ionization area near the collector will be affected and then measurement errors will occur.

7.2 Measurement of Electrostatic Charge Quantity

7.2.1 Direct Measurement Method of Charges

As shown in Fig. 7.7, a metal charged body insulated to the earth has a charge number of Q and capacitance to ground C_0, assuming that the ground potential is zero, then the potential of A is:

$$U_A = \frac{Q}{C_0} \tag{7.6}$$

If use the potentiometer with the input capacitance C_f to measure the voltage U_A, the above equation becomes:

$$U_A = \frac{Q}{C_0 + C_f} \tag{7.7}$$

As can be seen from Eq. (7.7), the voltage U_A is proportional to the charge quantity Q. After reading out the voltage, if the capacitor is known, the charge quantity Q can be calculated. But capacitance C_0 is changed with the changing measured object and environment, while C_f is inherent for an instrument. If the measured charged body's capacitance to ground C_0 is undetermined, we can increase the input capacitance C_f of the instrument to make $C_f >> C_0$, thereby the impact of C_0 is negligible and $Q = C_f U$ is considered to be right. Given the measurement of electrostatic charge quantity is conducted on the basis of the measurement of electrostatic voltage, voltage instruments are also required to have high input resistance to reduce measurement errors.

Fig. 7.7 Schematic of direct measure charges method

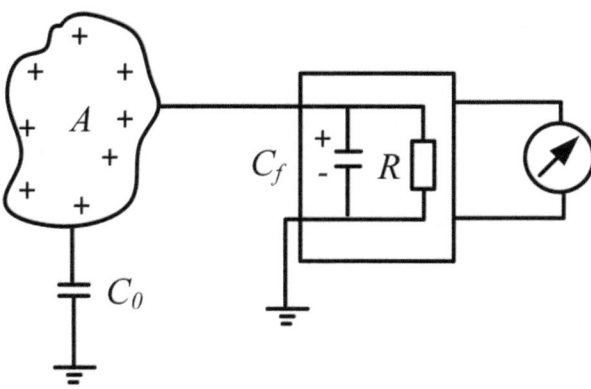

Fig. 7.8 Principle diagram of measuring charges By using Faraday cup. 1- outer cup; 2- inner cup; 3- capacitor; 4- electrostatic voltmeter; 5- insulated support

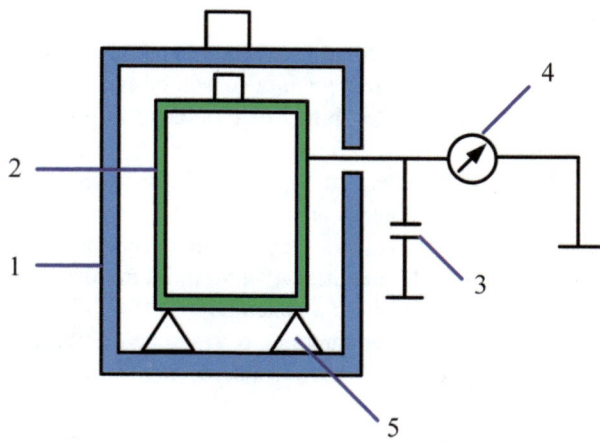

7.2.2   Measurement of Charge Quantity by Electrostatic Induction Method

The measurement of metal charged body is actually to transfer most charge Q on the charged body to the large capacitor C_f, measure the voltage of both ends of C_f and then reflects the value of Q. However, for electrified insulators, this transfer cannot be directly achieved. Therefore, the method of electrostatic induction can be utilized in virtue of the Faraday cup. Additionally, another limitation is that it cannot discern bipolar charging.

The measuring principle of the Faraday cup is as shown in Fig. 7.8.

During the measurement, put the charged body into the Faraday cup 2, so that charges with equivalent amount to charged body's charges and with the opposite symbol and the same symbol as charged body are induced on the inner and outer surfaces of the inner cup, respectively. In this case, the electrostatic voltage value U between the outer wall of the inner cup 2 and the grounded outer cup 1 is read by the electrostatic voltmeter 4, and the capacitance value C of the system (the sum of the capacitance between inner and outer cup, the input capacitance of voltmeter and the capacitance of the capacitor 3 in parallel with the device) is measured from the capacitance measuring instrument. Then the measured electrostatic charge quantity Q can be calculated according to the formula $Q = CU$.

7.3   Measurement of Electrostatic Properties of Powder

Powder is a special form of solid matter, having significantly different charged properties compared to other solid matters. Such a difference is caused when non-uniformity and dispersivity of powder together with random arrangement between the particles lead to non-uniformity, instability and anisotropy in its electrical properties. In addition, with a generally larger hygroscopicity, the measurement of the electrical properties of powders is greatly influenced by humidity. Besides, the measurement is also sometimes quite sensitive to temperature and pressure. All these combined factors result in poor reproducibility in the measurement of electrostatic properties of powder.

During the process of airflow processing and pipeline transportation, powder, due to frequent contact and separation between materials, pipe wall and wall of container as well as among powder material particles, tends to be charged. Moreover, some powder materials like ammonium nitrate and TNT explosives and other explosive products, with a general $10^{11} \sim 10^{15}$ Ω·cm specific resistance of volume, are within the danger range of being apt to accumulate static electricity. For this reason, people show more concerns about electrostatic protection of powders. Therefore, although with a poor reproducibility, the measurement of powder electrostatic parameter can still provide some quantitative descriptions and comparative data for the electrostatic properties of powder material, thus the research on the measurement of electrostatic properties of powder is still of practical significance.

7.3.1 Measurement of Specific Resistance of Powder

Specific resistance of powder refers to the resistance contained per unit volume of powder. Just as the importance of resistivity to solid medium, specific resistance of powder marks not only an important sign of insulation properties of powders, but also an important physical quantity reflecting the electrostatic properties of powders.

Usually, four methods are most commonly adopted in measuring specific resistance of powder: probe method (also known as sonde method), two-electrode method, three-electrode method and concentric-cylinder method. Basic principle used in these methods is Ohm's law. When taking measurement, the applied voltage is 100 V generally. For powders with a higher specific resistance, the applied voltage will be 250 V or 500 V. For low resistance powder with a specific resistance less than 10^5 Ω·cm, the measuring voltage will be within $5 \sim 50$ V. The measurement results of specific resistance of powder are decided by shapes, sizes and bulk densities of sample materials. Besides, humidity and the surface pollution status of containers or electrodes will also influence the measurement results greatly (Fig. 7.9).

7.3.1.1 Probe Method

The measuring process is as follows: adjust power voltage to make the voltage between needle electrode 1 and main electrode 3 approach the breakdown voltage of the powder sample, then record the voltage value U at this moment and current

Fig. 7.9 Principle diagram
of measuring powder
specific resistance by using
the needle electrode
method. 1- needle electrode;
2- Powder sample; 3- main
electrode; 4- Insulated
container; 5- Insulated
support

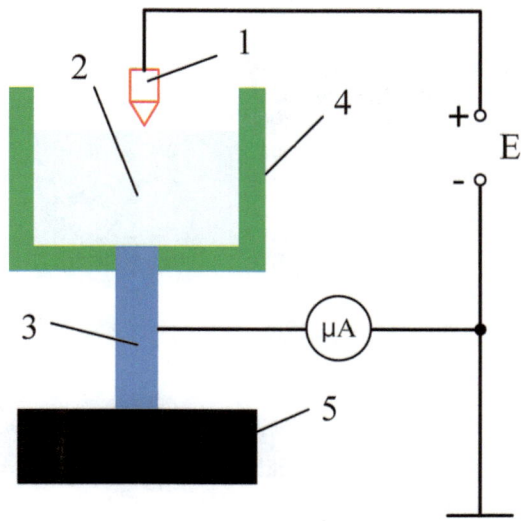

reading I on the micro-ammeter as well as volume resistance R_V ($R_V = U/I$) of the
powder material to be measured. Since R_V is inversely proportional to S (S refers to
the area of the main electrode 3) and directly proportional to the distance between
needle electrode and main electrode, the proportional coefficient between them is ρ
(the specific resistance of the powder), namely:

$$\rho = R_V \frac{S}{b} = \frac{U}{I} \frac{S}{b} \tag{7.8}$$

Wherein:

ρ specific resistance of the powder sample, $\Omega \cdot cm$;
R_V volume resistance of powder sample to be measured, Ω;
U added measuring voltage between needle electrode and main electrode, V;
I current on the micro-ammeter, A;
S area of main electrode, cm^2;
b distance between needle electrode and main electrode, cm;

   Probe method, due to adoption of structure of needle electrode, may cause large
error.

7.3.1.2   Two-Counter Electrode Method

Figure 7.10 shows the measuring principle of powder specific resistance by using
the two-counter electrode method.
   The measuring details are as follows: container 2 is made of highly insulating
materials, and both electrode A and B are square metal plates (generally

Fig. 7.10 Principle diagram of measuring powder specific resistance by using the Two-counter electrode method. 1- Powder sample; 2- Insulated container

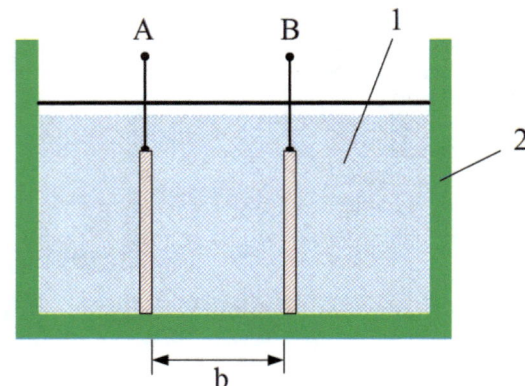

50 mm × 50 mm in dimension). b refers to distance between the two electrodes. Material sample 1 is placed in container 2. During measurement, if the voltage applied between the two-counter electrodes is U and current passing through two counter electrodes is I, then the volume resistance of the powder is R_V ($R_V = U/I$). Since the plate area is S and distance between the two electrodes is b, thus according to the Eq. (7.8), b is usually taken as 1 cm, substitute all the items into the formula, we can get:

$$\rho = 25 \cdot \frac{U}{I} \tag{7.9}$$

Wherein the unit of ρ is $\Omega \cdot cm$.

7.3.1.3 Three-Electrode Method

Figure.7.11 shows the measuring principle of powder specific resistance by using the Three-electrode. A, B and C are all metal concentric circular electrode plates, D_A and D_B refer to the outer diameters of electrode A and B respectively, and D_C refers to the inner diameter of electrode C. Electrode A, B and C are respectively a high-voltage electrode, a measuring electrode, and a guard electrode.

When measuring, a DC voltage U is applied between electrode A and B. If the current flowing through the measuring electrode B is I, the volume resistance powder will be R_V ($R_V = U/I$). If the guard electrode C is grounded, the current flowing through the inner wall of the container will be leaked, thus the measurement accuracy will be improved.

Since the specific resistance of the powder ρ equals $\rho = R_V \frac{S}{b}$, when calculating the effective area of measuring electrode, by taking the existence of guard electrodes into account, we can adopt effective radius r_0 ($r_0 = \frac{D_B}{2} + \frac{(D_C - D_B)}{4} = \frac{D_B + D_C}{4}$),

Fig. 7.11 Principle
diagram of measuring
powder specific resistance
by using the Three-
electrode. 1- Insulation
container; 2- Powder
sample; 3-Weight

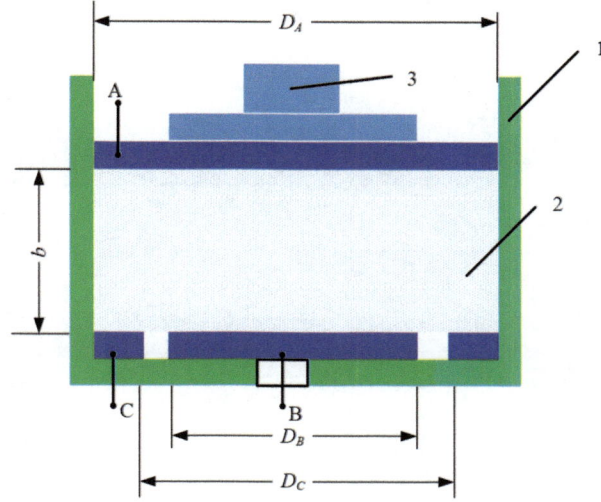

thus we can know that $S = \pi r_0^2 = \frac{\pi}{16}\left(D_B^2 + 2D_BD_c + D_C^2\right)$. After substituting
S and R_V into the formula, we can get:

$$\rho = \frac{\pi}{16}\left(D_B^2 + 2D_BD_c + D_C^2\right) \cdot \frac{U}{b \cdot I} \tag{7.10}$$

During the actual measurement, D_B, D_C and D_A are respectively taken as 2.5 cm,
3 cm and 4 cm, after substituting them into the formula, we can get:

$$\rho = 5.94\frac{U}{b \cdot I} \tag{7.11}$$

Three-electrode method, due to its high precision and good reproducibility in
measurement, is employed more.

7.3.2   Measurement of Electrostatic Potential (Voltage) of Powder

Measurement of electrostatic potential of powder, marking an important means to
reflect the charged status of powder, will provide quantitative data for electrostatic
protection of powders, making it gain more attention from the public. Figure 7.12
shows the measuring principle.

During measurement, put the metal probe 1 into the inside of powder pipeline,
and make sure good insulation between the metal probe and the inside of powder
pipeline by utilizing substances like polytetrafluoroethylene. Connect probe 1 with

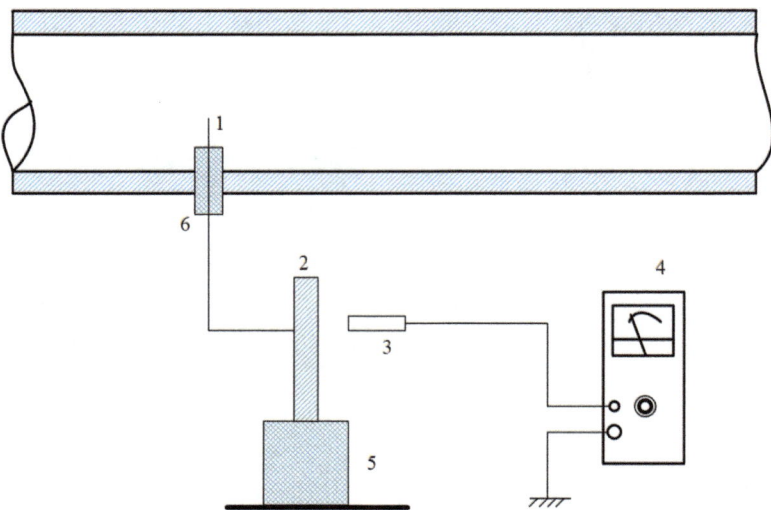

Fig. 7.12 Principle diagram of measuring powder electrostatic potential. *1* Measuring probe of metal, *2* collector plate, *3* Measuring probe of Instrument, *4* Electrostatic indicating instrument, *5* Insulation supporter, *6* Insulation

collector plate 2 through wires to make sure that collector plate and powder material in the pipeline are with equal potential. Measuring system composed of probe 3 and indicating instrument 4 can measure the potential of the collector plate.

7.3.2.1 Measurement of Specific Mass Charge Density of Powder

Mass charge density of powder refers to the amount of charge contained per unit powder. It is an important parameter to reflect the charged degree of powder. Once the material type is determined, the mass charge density of powder is related to the gas-solid ratio and the flow velocity of powder in the pipeline. If the mass charge density of the material exceeds the limit value, the surface electric field strength of each powder particle will be able to ionize the surrounding atmosphere, which will cause electrostatic discharge of powder. Therefore, the measurement of mass charge density of powder is very important for the prevention of electrostatic discharge accidents.

The principle and method in measuring mass charge density is: take the powder particle samples out of the pipeline and put them into the Faraday cup, and then measure the amount of charge carried in the powder samples, finally calculate the mass charge density of powder according to the definition of parameters and the formula $\rho_m = Q/m$. Usually, the unit of measurement of such a parameter is µC/kg.

In the actual measurement process, due to the weight increasing of materials caused by strong hygroscopicity of the powder and charge leakage and loss during operation, the measured values are often smaller than the actual value. One possible way to improve is to place a miniature Faraday cup directly in the pipeline, and use wire (the wire must be insulated from the pipe wall) to lead the Faraday cup out, then directly measure the electric quantity of the powder material by using the electrostatic voltmeter, and finally calculate the mass charge density.

7.4 Measurement of Electrostatic Charge Density of Liquid

For the measurement of charged stagnant liquid placed in the container, if the charge distribution is uniform, a Faraday cup can be used to measure the charge. Then, by dividing the charge measured in the previous step with the volume of the liquid to be measured, the electrostatic charge density can be achieved.

Taking the oil flow as an example, the measurement of the charge density of the flowing oil is actually carried out by measuring the potential. Oil in the pipeline is generally turbulent flow, charge distribution inside the pipe can be considered approximately uniform, thus the pipeline can be seen as an infinitely long uniform charged body, and the potential distribution on its cross section is shown as:

$$U = \frac{\rho r_0^2}{4\varepsilon}\left(1 - \frac{r^2}{r_0^2}\right) \tag{7.12}$$

Wherein:

U the potential on the section of the pipeline with an inner radius r, V;
ρ charge density, $\mu C \cdot m^{-3}$;
ε dielectric constant;
r the distance between any point on section of the pipeline and the center of the circle, m;
r_0 internal diameter of the pipeline, m;

From the above equation, we can know that U_r at r is proportional to ρ, making. $\beta = \frac{r_0^2}{4\varepsilon}\left(1 - \frac{r^2}{r_0^2}\right)$, thus $U_r = \beta\rho$.

Obviously, β is the function of r. At the center of the pipe, when r = 0, β can be regarded as β_0, thus $\beta_0 = \frac{r_0^2}{4\varepsilon}$.

If the potential at the center of pipe is U_0, then $U_0 = \beta_0\rho$, thus we know that:

$$\rho = \frac{U_0}{\beta_0} \tag{7.13}$$

Upon such a transformation, the measurement of ρ is transformed into the measurement of the potential at the center of pipe,

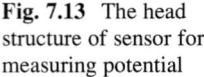

Fig. 7.13 The head
structure of sensor for
measuring potential

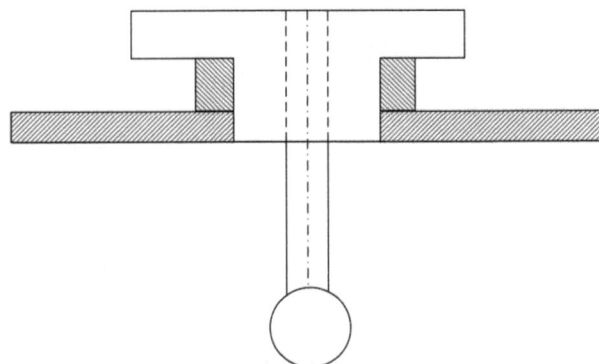

As shown in Fig. 7.13, the potential at the center is led by the pipe on the sensing head via a ball-and-pole electrode. The electrode is insulated from the metal pipeline. When the charged oil flows in the pipeline, the ball-and-pole is charged and the potential will rise gradually and will also leak to the pipeline wall simultaneously. When the charging current is equal to the leakage current, equilibrium will be achieved, and then a stable potential of the ball-and-pole will also be obtained.

Chapter 8
Electrostatic Utilization and Protection in Multiphase Flows

Through previous introductions, we have already got the idea that the process of contact and separation between any two objects will cause static electricity. Even if these two objects are made of the same material, due to the different conditions of their surfaces (such as surface contamination, corrosion and roughness), they will possess different work function, which will still make them produce static electricity during the process of contact and separation. Besides, through static induction or static polarization, discharged objects can be electrified. This phenomenon of objects being charged of static electricity could happen among solids, liquids, gases and powders. Therefore, the generation of static electricity is a very common natural phenomenon.

When static electricity exceeds a particular limit (which can be shown in forms of field strength, electric potential or stored energy) and be within a suitable objective environment, it will bring harm to production environment, products and human bodies in its different and unique ways.

The generation of static electricity is almost inevitable. However, its volume can be reduced to an acceptable level through various effective measures, so as to minimize its damage. Although the safeguards of static electricity that are applicable in production are multifarious, their basic ideas are always closely related to the following several points:

1. Minimize the generation of electrostatic charges;
2. Eliminate the already generated electrostatic charges as soon as possible, including accelerating their leakage, neutralizing and lowing their density;
3. Minimize the electrostatic hazards
4. Be strict with the management of electrostatic protection to ensure the effective implementation of protection measures.

© Springer Nature Singapore Pte Ltd. 2017
Z. Gu, W. Wei, *Electrification of Particulates in Industrial and Natural Multiphase flows*,
DOI 10.1007/978-981-10-3026-0_8

8.1   Main Contents of Electrostatic Protection

Various protection measures are integrated as

Approaches	Main contents	Concrete measures
1. To prevent the generation of static electricity	(1) Controlling the generation environment of static electricity	(1) Temperature control. Under the condition of not resulting in corrosion or rust of equipment or products, or any other damages, increase the temperature as possible;
		(2) Temperature control. If permitted, reduce the temperature as possible, including environment temperature and object-contact temperature;
		(3) Dust control. It is an important measure to prevent attachment (adsorption) electrification;
		(4) Electrostatic protection measures should be adopted during the transportation, storage, packing and unpacking of electrostatic sensitive products;
		(5) Control the velocity of spray, flow, transport and separate; apply moderators to the transporting pipes of liquids, powders, etc.
	(2) Requirements for material selection	(1) Precedence of materials that are definitely or likely to be contacted and separated should be as close as possible on electrification sequence list;
		(2) Surfaces of materials should be smooth and clean;
		(3) Use static conductive materials and static dissipating materials.
	(3) Technological control measures	(1) Develop and implement anti-static operating procedures;
		(2) Utilize anti-static tools (soldering iron, tin extractors, etc.);
		(3) Adopt anti-static package;
		(4) Set necessary standing time for liquid materials with possibility of electrostatic combustion and explosion;
		(5) Minimize the contact pressure, time and area of objects, and limit the velocity of movement.

(continued)

Approaches	Main contents	Concrete measures
2. To reduce and eliminate electrostatic charges	(1) Grounding	(1) Worktables, chairs, countertops and table pads are properly grounded to the floor;
		(2) Human body grounding;
		(3) Tool grounding;
		(4) Equipment and instrument grounding;
		(5) Pipelines, transportation facilities, canning equipment, storage equipment (instrument) grounding.
	(2) Humidification	(1) Utilize any proper humidifiers and sprayer units;
		(2) Wipe the floor with wet mops or water the floor to increase the temperature of environment or temperature near charged bodies;
		(3) Select hygroscopic materials if it is allowed.
	(3) Neutralization	Select preferable types of electrical static eliminators according to workplaces and shapes and characteristics of charged objects to eliminate electrostatic charges on appliances, equipment, products, workplaces, instrument and human bodies.
	(4) Adulteration	(1) Attach a layer of material on the surfaces of non-conductive materials and appliances through ways of spraying, smearing, plating, coating, printing and pasting, etc. to increase the surface conductivity and accelerate the leakage and release of charges;
		(2) Mix non-conductive materials such as plastic, rubber and anti-corrosive paint with metal powder, conductive fibers, black carbon powder, etc. to increase their conductivity;
		(3) Add chemicals as anti-static additives into easy-to-generate-static-electricity liquids to improve conductivity of liquid materials.

(continued)

Approaches	Main contents	Concrete measures
3. To reduce electrostatic hazards	(1) Adopting static shielding and grounding design	(1) For static sensitive parts and components, adopt static shielding measures of adding protective caps, covers and shields to reduce mechanics, responses and discharging hazards of static electricity;
		(2) Avoid isolated conductors as possible;
		(3) Set up shielding monitoring wells in liquid storage tanks, etc. to ensure the safety of sampling and testing.
	(2) Ensuring the electrostatic safety of equipment, instrument and workplaces	(1) Control the combustible and explosive liquids or powders to make the concentration of explosive compound under the limit concentration of combustion and explosion;
		(2) Keep grounding facilities and systems in workplaces correctly and effectively grounded;
		(3) Make electrostatic potential of points in operation areas within acceptable limits;
		(4) Install local dischargers and discharging brushes to release electrostatic energy through corona discharge, in order to make the accumulated energy in the safety range;
		(5) Strictly implement operation procedures of electrostatic protection.
4. To be strict with the anti-static managements	(1) Establishing and improving responsibility system and rules and regulations	(1) Establish and improve management responsibilities of electrostatic protection for all staff members with examination system;
		(2) Establish analysis system for electrostatic accidents;
		(3) Establish concrete and detailed anti-static operating procedures;
		(4) Establish regular test system for anti-static performances of equipment, facilities, instrument, tools, etc.;
		(5) Establish analysis system for electrostatic damages of products.
	(2) Training and education	(1) For staff members of different levels and positions, such as leaders, managerial and technical personnel and operating crews, implement educations of anti-static awareness,

(continued)

Approaches	Main contents	Concrete measures
		knowledge, techniques and safety requirements;
		(2) Train and examine the operating skills of operating crews of different positions.
	(3) Utilization of warning devices, labels and signs	(1) Tag or mark the static sensitive products on products themselves and their internal and external packages;
		(2) Tag or mark the static sensitive components and parts of facilities and equipment with warning signs as standards require;
		(3) Prescribed special marks should be applied to anti-static operational fields;
		(4) Warning devices should be installed on key control locations to remind people of it in time.
	(4) Conducting inspections and tests according to standards and specifications	(1) Measure and monitor environment parameters of specific requirements (such as humidity, temperature, concentration and electrostatic potential) according to regulations;
		(2) Check the grounding conditions of human bodies, equipment, facilities and systems as regulations require;
		(3) According to standard regulations, test the static sensitivity of products, and establish quality analysis and feedback system.

8.2   Electrostatic Protection of Powders

Powder materials in motion, such as fast flowing, shaking or vibrating, will rub against, collide with and separate from transporting pipelines, vessel walls and conveyer belts; also, powder particles themselves will mutually rub, collide, fracture, break and separate. All of these phenomena will generate static electricity, which contribute to the main mechanism of powder electrification. These phenomena are inevitable in manufacturing procedures of industrial powders, and can be encountered at any time. Therefore, powder electrostatic protection is a main aspect of industrial electrostatic protection.

1. Internal factors of static electricity generated in powders

   According to the double layer theory of static electricity generation, the occurrence of static electricity is closely related to the conductivity of objects themselves. If electrostatic conductivity is represented by electric volume resistivity ρ_V, then when $\rho_V < 10^{10}$ Ω·cm, it is of good conductive property, and thus electrostatic charges generated by mutual fiction are hard to accumulate due to rapid distribution, diffusion and leakage; when $\rho_V > 10^{16}$ Ω·cm, due to the insufficiency of easy-flowing free charges, and thus it is hard to transfer charges and form double-electric layer to generate static electricity; only when ρ_V ranges from $10^{11} \sim 10^{15}$ Ω·cm can static electricity be generated in the easiest condition.

2. External factors of static electricity generated in powders

Solids will carry electrostatic charges due to friction, roll extrusion, squeeze, detachment, or attachment by charged objects. The volume of electrostatic charges not only depends on the internal characteristics of solids, but also the external factors such as friction velocity, friction pressure, friction contact area and environmental humidity.

Obviously, only by taking targeted protection measures from aforementioned both internal and external aspects can static electricity of solids be reduced and eliminated effectively.

8.2.1 General Measures of Powder Electrostatic Protection

1. Select and use metal or conductive raw materials as possible to fabricate various processing apparatus, pipelines and containers, which also must be grounded.
2. If it is permitted, increase the relative humidity of operating environment to reach at least 65%, or add proper amount of water into materials to make the conductivity of non-electrostatic-conductive materials at least 10^{-12} S·m^{-1}.
3. Apply electrical static eliminators. For combustible powders, apply the explosion-proof type with external power supply. Radioactive eliminators can be applied when operators are adequately protected.

8.2.2 Electrostatic Protection Measures in each Process of Powder Processing

1. Electrostatic protection of transporting powders through pipelines

Static electricity of powders is mostly generated through friction and collision between powder materials and pipelines. During the pneumatic transport of powders, if other conditions are fixed, then the charging current density J is in proportion to 1.8 power of powder airflow velocity v, and inversely proportional to equivalent diameter of powder particles d. Therefore, to reduce static electricity is to reduce and maintain the airflow velocity, of which the main measures are: diameters of transport pipelines should be as large as possible to reduce the velocity as possible; velocities should be limited, the smaller the powder particle size is, the slower the speed is, specific velocity should depend on material types and air humidity, etc., and be adjusted to the allowed range through voltage measurement; pipelines should be unobstructed, it is not allowed to set obstructions like grids, bends and contractions of pipelines should be less to prevent disorders and rapid changes of transportation, meanwhile, inner walls of pipelines and hoppers should be cleaned regularly to prevent accumulation of powder materials; pipelines, hoppers, funnels should be made of conductive materials as possible, or materials matching the powders (i.e. close to each other on electrostatic sequence list), and the conductive materials should be grounded correctly and firmly; moderators can be applied to hoppers and bunkers to make charged powders leak most (at around 95%) of their charges in moderators.

2. Electrostatic protection of transporting powders through conveyer belts

Shape of helical blades and screw revolution of worm conveyers all affect the generation of static electricity, thus the screw revolution should be adjusted according to tests to make the volume of static electricity in a safe range; abnormal vibration of conveyer belts should be avoided, meanwhile, powder splash should be prevented through multiple ways such as controlling the transport velocity; stability of transporting should be maintained, accumulation on belts should be prevented, surfaces of belts should be cleaned and examined regularly; conveyer belts should be made of conductive materials as possible, and be electrostatic grounded by rollers, etc..

3. Electrostatic protection of transporting powders through hoppers

Wall slope of hoppers and funnels should be approximately vertical to reduce the friction area as possible; hopper walls should be smooth to prevent the declining resistance and friction of materials from increasing; hopper and funnels should be made of conductive materials as possible, and be grounded; and materials of hoppers should be selected according to parameters of powder materials such as type and size to minimize the generation of static electricity.

4. Powder electrostatic protection during handling, transporting and blending

During the process of handling, transporting, feeding and bagging of powder materials, when feeding from bags to containers through belts, hoppers or funnels, it should be as continuous and quantitative as possible, and should avoid transporting or inputting large amounts of powder materials in one shot at high speed. While taking out powder material from bags, the strong shaking of bags should be avoided. Impurities mixing in powders should be avoided. Once impurities are found, they should be immediately picked out.

5. Electrostatic protection for during crushing and sorting

Powders of specially uneven shapes should be smashed into adequate particles first, then shattered slowly to prevent intense shattering; any impurities mixed in powders should be eliminated; speed of crushing and sorting should be as slow as possible, dealing a small amount of powders each time, when processing, abnormal temperature rise should be avoided; material type permitting, water should be added to increase humidity; griddles should match powder materials to minimize static electricity; devices like griddles and sorters should be made of conductive materials and grounded.

6. Powder electrostatic protection during collecting process

To minimize the generation of static electricity, the material of filters should select sack, better mixed with conductive fibers to match with powder materials; the area of sack filters should be big enough, and not locally collect massive powder materials; when applying cyclone separators, the wind velocity should be as small as possible, besides, the inner walls of cyclone separators should not have any projections, in order to prevent the local accumulation of powder materials, the cyclone separators should be cleaned regularly.

7. Powder electrostatic protection during drying and spraying process

The flow velocity of powders should not be too large, and be adjusted according to the type and quantity of powders along with air quantity and air temperature, besides, rapid changes of air quantity should be avoided; the arid condition should not be excessive while the air temperature should not be too high; selection of ejection pressure, air quantity and shape of outlet should minimize the spraying velocity; distance between sprayer and spray objects should be as long as possible to reduce collision potential stress.

8.3   Electrostatic Protection of Liquids

Liquid electrostatic protection is an important part of electrostatic protection. Being security-focused, it prevents accidents caused by liquid static electricity such as fire and explosion.

Widely used industrial liquids, especially light hydrocarbon oils such as gasoline, industrial kerosene and diesel oil, deserve special attention in industrial electrostatic protection due to their characteristics of high fugacity and low flash point and with resistivity in a dangerous range of $10^{11} \sim 10^{15}$ $\Omega \cdot$cm.

From previous introduction it can be inferred that liquid static electricity can be generated through almost every industrial process and procedure; it could be the devices that carry the charges, or liquids themselves.

8.3.1 General Measures of Liquid Electrostatic Protection

1. Grounding

Transportation pipelines, machining and storage facilities and appurtenances of liquids should be grounded. Metal conductors should be directly grounded while static conductors and semi-conductors are indirectly grounded; to leak charges through grounding, the relevant pipes and devices should be made of metal materials as possible, the general bleeder resistance between metal conductors and the earth should be less than 1 MΩ under normal conditions.

2. Limit the flow velocity of liquids

In section 3.1.3 of Chapter 3, it mentioned the calculation equation of saturated streaming current during liquid streaming electrification as $I_\infty = Av^\alpha d_p^\beta$, which indicated that the current of streaming electrification mainly depends on the internal diameter of pipelines and flow velocity, therefore, to prevent generating excessive streaming electrification current, it is necessary to limit the initial velocity and maximum velocity of liquids.

No matter what kind of liquid media are filled and transported in what methods through what pipelines and containers, when the filling process starts, the flow velocity of fluids transported inside the pipelines should be limited within 1 m/s to ensure the security. Increasing of velocity is only allowed under following conditions: when filling with ducts, the outlet is completely immersed into the liquid; when filling through the lateral bottom of the tank, liquid is at least a caliber higher than the outlet; when the tank is internal floating roof tank, the floating roof is completely surfacing the fluid; remaining oil, water and air inside the pipelines is completely exhausted; oil level of the cabin rises beyond the stringer of cabin bottom and water stabilizes at the bottom of the oil tanker.

3. Apply electrical static eliminators

When electrostatic accumulation cannot be reduced through flow velocity control and other methods, in order to reduce the static electricity of hydrocarbon fluids such as gasoline, electrical static eliminators for oil pipelines can be applied near the pipe outlet.

The traditional pipe electrostatic-eliminator included oil conveying inner tube and a grounded outer tube rounded cross-section discharge needles was fixed on the outer tube, and got through the inner tube. When the charged oil flowed into the tube, the capacitance generated by inner and outer tubes raised the potential of oil. So, the high potential difference generated between oil and grounded discharge needles. In needle tip area nearby, under the function of the strong electric field generated by the high potential difference, the charge in oil would be neutralized by the heterocharge inducted out by grounding, so as to achieve the purpose of eliminate electrostatics in oil (Zheng et al. 2013).

4. Apply moderating devices

Large quantities of electrostatic charges are generated when hydrocarbon oils pass through meticulous filters, thus a relaxation time of at least 30s should be reserved from the outlets of filters to containers.

Attenuation characteristics of liquid static electricity depend on its conductivity σ and dielectric coefficient ε, thus when designing the liquid dwelling time in moderators, the type of liquid transported in pipeline should be considered, normally, the relaxation stay is 3 times of the liquid relaxation time.

5. Improve the environmental conditions around charged bodies

When the mixture of hydrocarbon liquid vapour and air approaches the limit of explosion concentration, ventilation measures of operational fields must be reinforced to reduce the concentration of the mixture, if necessary, inert gas system should be equipped.

8.3.2 Technological Protection Measures for Various Devices

1. Electrostatic protection for transportation pipelines

In order to reduce the charges generated by streaming electrification of liquids inside the pipeline, bends and throats of pipeline should be as less as possible. Transporting pipelines are better made of conductive materials and grounded themselves, when using the non-conductive materials, measures such as internal lining wire gauze should be adopted, besides, attention should be paid to electrostatic continuity nearing the junctions. All metal components of pipelines, including metal cladding of sheathings must be grounded. Materials of pipelines are preferable to that generate lesser static electricity between transported materials. Under the condition of maintaining the same capacity of transporting liquids, the increasing of diameters of pipes or hoses should be considered to reduce flow velocity. Oils transported through pipelines should avoid being mixed with impurities such as air, water and dust as possible, particularly, the infiltrating of dissolve gels like rubber and asphalt should be controlled. When there are obvious a second phase inside the pipelines, the flow velocity must be reduced to less than 1 m·s^{-1}.

2. Electrostatic protection for oil tankers and ships

Before loading and unloading oils, ship bodies must be grounded to the earth grounding terminals. Electrical continuity of hoses should be examined before using them to transport light oils. Initial velocity of oil loading shall not be larger than 1 m·s^{-1}, only when the outlet is immersed can the flow velocity be increase. When cleaning the tanks, anti-static operating procedures must be observed. When the tanks are loaded with oils whose flash point is 60 °C, it is preferable to equip the tanks with inert gas systems.

3. Electrostatic protection for stirring, mixing and blending equipment

All metal components of stirring, mixing and blending equipment should be connected to each other and grounded. If the equipment has insulating linings, internal grounding measures can be adopted to accelerate charge leakage. It is prohibited to blend gasoline, kerosene, or light diesel oil with compressed air. When blending heavy diesel oil and so forth with compressed air, the wind pressure shall be controlled under 343 kPa, and the blending temperature of oil must be 20 °C lesser than oil flash point. If jet mixing is adopted, as long as the fluid column won't erupt over the liquid surface and the metal components of equipment are grounded, it is unlikely to cause electrostatic ignition.

Reference

Zheng, W., Y. Hu, C. Liu, X. Wei, and L. Hao. 2013. The simulation of electrostatics elimination effect influenced by the cross-section shape of discharge needles. *Journal of Physics: Conference Series, IOP Publishing* 418 (1): 012017.

Chapter 9
Potential Applications of Particulate Electrification

The basic charge structure of the convective updraft in a thundercloud has four charged regions alternating in polarity, with the lowest being positively charged. The three lower charged regions fit well with the classical tripole charge model of thunderstorms, which suggests a main negatively charged region between a main upper positively charged region and a lower positively charged region. The formation of the tripole charged structure is usually explained by the ice crystal-graupel collision charging which has been commonly recognized as the dominant mechanism of severe thunderstorm electrification. The relative growth rate was based on laboratory experiments and this hypothesis can indeed explain the phenomenon of charge-reversal and the formed tripole structure of thunderclouds, but the mechanism of the relative growth rate is still unclear.

In many current electrification models, the updraft-downdraft transition layer is regarded as a separation zone of charged particles, irrespective of the convection leading to non-thermal equilibrium of particles. In fact, there is a strong convection in thunderclouds, in which the near-surface warm updraft would mix with the bottom of the clouds, to form a mixed-phase region of water droplets and ice crystals in each thundercloud. Liquid/solid particles are dragged by the updraft and downdraft in the clouds, leading to temperature differences between particles and their environment. Based on the non-thermal equilibrium ionization of cloud particles and the consequent growth of the electric potential difference between the surface and inside of the cloud particles, it's expected to propose a general dynamic framework mechanism for thundercloud electrification.

The ascent of air in developing thunderclouds could enhance the formation and growth of hydrometeors and the descent of air would cause the dissipation of hydrometeors. The hydrometeors are present in non-thermal equilibrium states due to the variation of vapour pressure in the dynamical and thermodynamic environments. The characteristic time, tr, for vapour diffusion for a crystal surface to respond to a changed local environment can be expressed as, where r is the crystal radius and DV is the diffusion coefficient for water vapour. For small ice

Z. Gu, W. Wei, *Electrification of Particulates in Industrial and Natural Multiphase flows*,
DOI 10.1007/978-981-10-3026-0_9

particle, the response time is in the order of microseconds. In contrast, it takes tens of milliseconds to reach thermal equilibrium in a new environment.

A higher growth rate or freezing rate would correspond to a higher temperature gradient. An estimate of the temperature gradient is the temperature difference between the particle and the environment divided by the perimeter of the particle, viz.: Here, the temperature difference would also manifest the temperature difference inside ice particle due to the slow heat conduction between the particle surface and its inner part in the new environment. Driven by the temperature gradient, an ionic migration inside the particle would be induced and the mobility difference between H+ and OH- would further make a particle to gain a net charge on its surface, although each particle should be neutral as a whole. The difference of surface ion concentration between different size particles would make particles acquire different net charges, negative or positive, after their collision/contact and separation. In addition, the growing states of cloud particles, freezing, riming, evaporating or sublimating, determines the charge sign of particle surface. In fact, the cloud particles in different layer of thunderstorms are in different state, which may be the cause for the multiple charge structure in thunderstorm.